Biomimicry

Nature as Designer

Written and illustrated by the 7th grade students of High Tech Middle School, San Diego, CA, in the Fall of 2015.

Edited by Brooke Young, Roxie Brunsting, and Ben Krueger.

For information please contact:
bkrueger@hightechhigh.org, bryoung@hightechhigh.org, or rbrunsting@ hightechhigh.org

Biomimicry

Nature as Designer

Inside, you will find detailed descriptions of product prototype ideas based on the existing principals of nature; known as biomimicry. Biomimicry is the process of copying an adaptation from nature, and creating a product or process based off of the adaptation. We have created products that can make the world a better place, ranging from cars to water pumps to inflatable cell phone covers to everything in between. We hope you find inspiration within!

-Molly Zucchet and Shreena Bhakta, Inventors

Table of Contents

BambO$_2$

By Collette McCurdy

Did you know that Bamboo is the fastest growing plant on Earth? Or that the Silver Maple is one of the trees that takes in the most carbon dioxide (CO2)? Or that every time you burn something, more carbon dioxide is released into the atmosphere? And since so many things are burned every day, there's a lot of CO2 out there. That much CO2 creates global warming. Which is exactly what my product is going to solve.

Most plants grow by creating more and more cells that stack up, making the plant grow in both height and width. But Bamboo grows differently from most seedlings. Instead, Bamboo plants only grow a couple of cells that stretch out really far (Tim Laughlin). Stretching out is a lot easier than producing more and more cells. Silver Maple trees are one of the trees that take in the most carbon dioxide.

 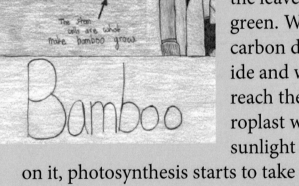

Trees have pores in their leaves called "stomata". These pores are what take in carbon dioxide and release oxygen. Inside the leaves, you'll find chloroplast. Chloroplast is what makes the leaves green. When carbon dioxide and water reach the chloroplast while sunlight shines on it, photosynthesis starts to take place. And when carbon dioxide and water are hit by that light energy, it results in turning into oxygen (Elearnin).

You might be wondering, what is global warming? Basically, it's the fact that our planet is gradually becoming warmer and scientists think that humans are responsible. We humans get most of our energy from burning fossil fuels. These fossil fuels release CO2 and other greenhouse gasses into the atmosphere. Yes, CO2. It's what plants need to survive, right? But if the world

I wanted to do something that would help the world in the best possible way.

has too much, it might even be harder to live, since the greenhouse gases trap the heat coming from sunlight. With the heat getting stuck in the atmosphere, the earth will become warmer and warmer, making life difficult not only for us, but for other living creatures, too. For example, glaciers and the polar ice caps will melt (save the polar bears!), oceans will become to warm (save the fishes!), causing major hurricanes (save humans!). Plants and animals won't be able to survive because of the huge temperature change (The Down - to - Earth Guide to Global Warming). This is why we need to do something about it. Now.

I am going to use both the Silver Maple tree and the Bamboo plant to create a prototype that will help prevent global warming. It's called the BambO2, because I am improving bamboo's ability to turn carbon dioxide into oxygen. The cells in the stem of the Bamboo plant will

be transferred into the trunk of the Silver Maple tree. With many Bamboo stem cells, the Silver Maple should grow quickly but still take in lots of carbon dioxide. When the Silver Maple grows quickly, it should get more leaves quickly, and be able to take in more carbon dioxide before most other plants. Also, when the Silver Maple grows quickly, it will grow seeds faster than usual. That means you will be growing more and more Silver Maple trees, too. Hopefully, there will be farms growing the trees on river bottom soils, which is where they will grow best.

The reason I chose to put the Bamboo plant and the Silver Maple tree together for my product is because I wanted to do something that would help the world in the best possible way. I knew trees collected carbon dioxide, and I knew I wanted to make something to help stop global warming. So I put them together, and Voila! a solution to global warming! Without a planet, humans wouldn't exist. That's why we need to do our very best to keep Earth at the right temperature so that we (and everyone else living here) can stay alive.

The stem cells will be transferred into the trunk of the Silver Maple

This will make the trees grow faster

Camel Water Pump Bins
by Shreena Bhakta

Have you ever wondered what life would be like if you didn't have access to clean drinking water? Unfortunately, this is many people's realities. By installing my prototype, water pump bins, in villages all over the world, it will help people with their education, hunger, health, and bringing people out of poverty.

Dromedary Camels are a type of camel that were tamed in the Arabian Peninsula. The other type of camel is called a Bactrian Camel. An easy way to tell the difference between the two is that the Bactrian Camel has two humps, and the Dromedary Camel has one hump. Dromedary Camels are often used for transporting heavy loads in caravans (National Geographic). Because of this, they have to have certain adaptations that are very useful for living in a harsh, dry environment. Their kidneys and intestines can conserve water well according to to the fact sheet on The San Diego Zoo's website. Oval shaped red blood cells reduce the risk of dehydration, as stated by Animal Corner, a British website. One adaptation that the camel uses is how it cools the blood going to the brain. The blood flowing to the brain goes through a special network known as the rete mirabile. It is a well-advanced network located at the low part of the brain (Cab Direct, Mathur, R. Jain, R. K. Nagpal, S. K.). As shown in the picture, the blood flowing in the carotid artery moves into the rete mirabile, where is gets cooled down using countercurrent heat exchange. It then moves up into the brain.

Like a normal water pump, water is pushed through a suction water line. It goes through multiple filters, which are located in a durable, plastic bin. The filters remove any dirt or bacteria in the water, and then

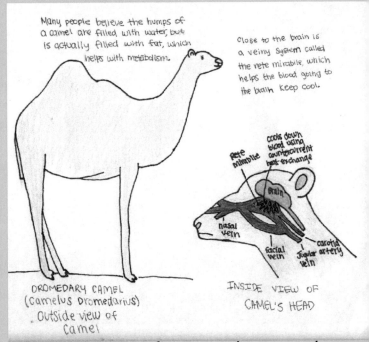

DROMEDARY CAMEL (camelus Dromedarius) Outside view of camel

INSIDE VIEW OF CAMEL'S HEAD

the water goes out of a water outlet. To start the water suction line, the villager must push the force rod/lever back and forth. The pipes in the ground connect to all other water pump bins in the village. The pipes have insulation to keep the water traveling in the bins keep

> *"The rete mirabile in the Dromedary Camel is an extraordinary adaptation that helps the brain keep algid in such blazing temperatures."*

cool. The pipe structure mimics the rete mirabile in the camel, the rete mirabile acts as "pipes for the blood" to be cooled down. The insulation mimics the camels outer fur, which can cover and coat the camel during cold and hot temperatures, using countercurrent heat exchange (CCMI Staff, Cashmere and Camel Hair Manufacturers Institute).

Water affects people's lives in so many ways. It is crucial for us. Water can help villagers in four different areas in their lives. Education helps students stay in school if they spend less time retrieving water and if they have secure and appropriate bathrooms. People's health can be reduced by having clean water minimizing sickness times so they can go to work and bring themselves out of poverty. By having access to clean water, there is less crop loss, which leads to less hunger. By having clean water, you can live without poverty. (Lisa McAllister, The Water Project) Normal water pumps have a bin in the in the ground that stores water, in which it can be pumped out. My invention uses ground water instead of water in the bin. Ground water can easily be found after monsoon season, a heavy rain that often floods villages. The ground can soak up most of that water and can be later used for the water pump bins. The pipes with insulation help keep the water from the ground cool while going to each and every water bin in the village, similar to the Dromedary Camel's rete mirabile. A normal water pump has

"Water affects people's lives in so many ways."

a pipe coming straight out of the bin full of water, and exits through a water outlet when a villager pushes back and forth the force rod/lever.

The rete mirabile in the Dromedary Camel is an extraordinary adaptation that helps the brain keep algid in such blazing temperatures. The pipes in the water pump bins keeps the water frigid while going to each and every water pump bin in the village. People should have access to water everywhere, and by installing water pump bins in villages all over the world, villagers will have an easier and better daily life.

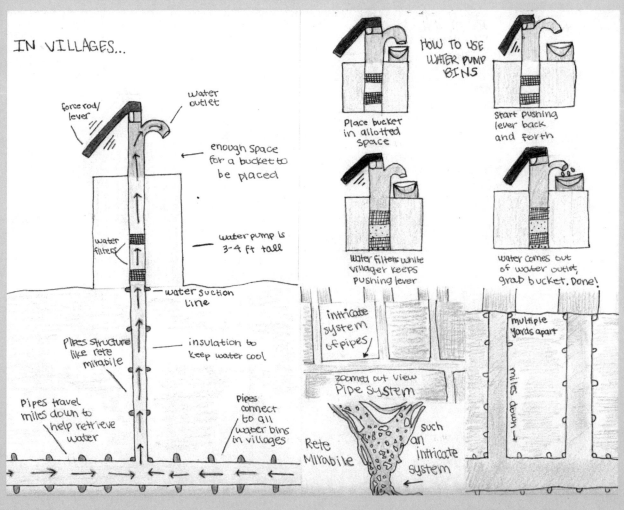

The Pufferfish Phone Case

by Molly Zucchet

"A lower priced innovative solution to protective phone cases."

42% of iPhone users have a cracked phone screen. And that's just iPhone users, not even counting the other 104 mobile phone brands (Trenholm, Richard. "Quarter of IPhones Have a Broken Screen, Says New Poll." CNET. N.p., n.d. Web. 01 Dec. 2015). With the Pufferfish Phone Case, smartphone users will no longer have to worry about spending unnecessary amounts of money on large bulky phone cases, and will have a slim, advanced, protective case they can rely on.

Pufferfish, also known as blowfish, are clumsy swimmers that are able to fill their flexible stomachs with tremendous amounts of water (or air) and expand themselves up to three times their normal size. From the moment they hatch, pufferfish are equipped with the ability to inflate, although the fish can't puff up to its full potential until adulthood. When alarmed, the pufferfish must widen its mouth by unhinging its jaws, in order to suck up 35 large gulps of water in a few short seconds. Thanks to the widening of the mouth, all the water (or air, if the fish is on land) is not in the puffer's stomach yet, it is still in it's mouth. Once all the water has been sucked in, and the pufferfish still feels that it is being threatened, it moves to the inflation stage. The puffer's stomach is extremely stretchy, and it has no rib cage to get in the way. The pufferfish "coughs" the water to the front of it's mouth, then down an open esophagus valve and into the stomach. As the stomach inflates the pufferfish becomes up to three times its normal size. Before the pufferfish inflated, it was already incapable of outrunning (or outswimming) its enemies. But because the pufferfish has expanded so much, it has lost even more of its mobility. Once the puffer is confident the threat is gone, it becomes

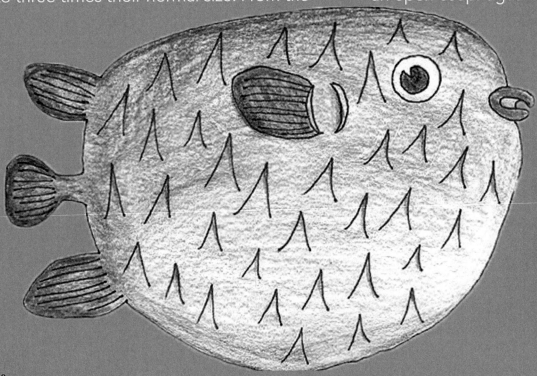

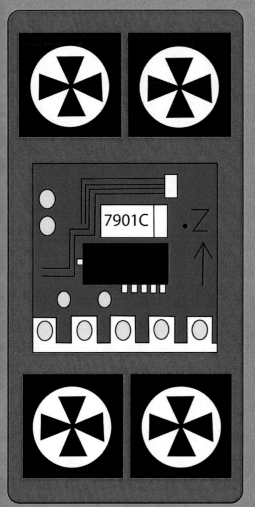

calm and forces all the water out of it's stomach. If too much air is inside the stomach, it can prevent the pufferfish from releasing the water, which can end its life (Ratliff, Dondi. "What Happens When Pufferfish Puff Up?" Animals. N.p., n.d. Web. 01 Dec. 2015).

Like the pufferfish, the Pufferfish Phone Case can expand with air when the phone is dropped. The phone case has an accelerometer inside of it, enabling the phone to sense movement in the case. An accelerometer is a small electronic inside your smartphone, made up of motion sensors. The motion sensors enable the phone to sense movement, rotation, and even earthquakes. Accelerometers are also used in prosthetic limbs, or other devices that use a lot of movement (Goodrich, Ryan. "Accelerometers: What They Are & How They Work." LiveScience.com. N.p., n.d. Web. 01 Dec. 2015). The case is made of thin plastic,

covered in strong flexible Bamboo Fabric, allowing the large amounts of air to sit between the strong fabric and the plastic. Bamboo fabric is similar to silk and cotton, though more practical. It is also wrinkle free, which will prevent the case from looking worn out after every time it drops. Inside of the case, there will be four 30 millimeter by 30 millimeter fans to inflate the case. The fans wheigh only eight grams, adding barely any extra weight or bulkiness to the case (VGA Card Cooling Fan sunsky-online.com. N.p., n.d. Web. 01 Dec. 2015).

The Pufferfish Phone Case is a lower priced innovative solution to protective phone cases. Leading protective brands such as OtterBox add an additional 30 millimeters to Apple's iPhone 6, whereas the pufferfish phone case only adds 10 millimeters. The pufferfish phone case is 1/3 of the OtterBox case, with over 15 millimeters in difference. Therefor your protective case options are a 30 millimeter case that adds unnecessary additional weight and bulkiness to a phone that is designed to feel slim and weightless in your hand, or a 10 millimeter case equipped with modern technology modeled after a unique fish with unique adaptations and a slim design to fit easily in any purse or pocket.

So I ask you... What's in *your* pocket?

What would you think or do if you get lost in caves, forests or even anywhere? You would stay in one place and wait hours and hours for someone to come and rescue you. Let's first talk about the problems with caves. Rocks tend to shift for some certain reason, causing you to separate from your group not knowing what to do. Now in this situation, I thought about designing a robotic bat to fly around different areas including the cave. Bats are split up into two different types, the megabats and the microbats. For the cave situation, you would need microbats to crawl or fly past little crevices to look for you and your group. What you are reading will convince you that robotic bats are more than a toy in life.

Bats can be cute and at the same time scary like you have seen in movies with vampires sucking blood from terrorized kids. Their strong feet allow them to hang over upside down making it look adorable. "70% of bats consume insects, sharing a large part of natural pest control. There are also fruit-eating bats; nectar-eating bats; carnivorous bats that prey on small mammals, birds, lizards and frogs; fish-eating bats, and perhaps most famously, the blood-sucking vampire bats of South America," says Defenders of Wildlife. It was also said they live all over the world but in the colder regions.

Echolocation Key
Blue Echo - Echo from bat
Red Echo - Echo from buterfly

They have five different fingers, wrists, elbows, two legs, knees and two ears as we do but they are far cooler than we are. They have other characteristics that are alike and unlike us, with even more to find.

With the robot bat, it would have the characteristics of the actual bat making it look similar. The robot head would have high quality speakers to speak and to use echolocation to see objects like people, trees, rocks and so on. This is how it works, the bat shoots out a sonar that spreads around. Then the bat would wait until the sonar comes back or not. If it does, the bat can determine how far their prey or object is by seeing how long the sonar would come back to it. Now the wings, they are made up of a certain material that is sturdy enough to resist bumping into obstacles. Sailcloth will make the robot bat glide like a mini hang glider. Other features on the wing would include expandable wings to glide faster and small metal fingers to make the bot look re-

alistic. Toes and tails will also be included.

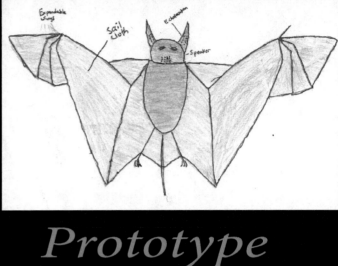

Prototype

Bats would be an important animal in life because of their amazing sense of smell and echolocation. They could be the perfect solution to getting lost, military situations, agriculture and exploring caves. Getting a little bit deeper, robot bats can save the day if any disasters occur like an earthquake or an avalanche.

Why bats you may ask. Well, for several of my opinions they look cool, they have wings, they suck blood, and they would be a great pet. For the scientific reasons, they use echolocation and have a great sense of smell. They may even save your life.

Air Ventilation

by Donnie Sullivan

More than half of all people in the U.S have air conditioning, showing two thirds of all homes in the U.S have air conditioning, which is around one hundred nine million people have air conditioning, out of two hundred eighteen million people (CsMonitor). Americans spend over 22 billion dollars a year combined on air conditioning, that's way too much money being spent every year on just air conditioning! The average cost of air conditioning is $2,800 for an entire household, think how much it would cost if you added it to a sky scraper of 100 plus floors. Some people don't have $2,200 just lying around to spend on air conditioning, so It's pretty expensive for a lot of people. I am planning on creating an air ventilation system which will be incorporated in households, and buildings. This ventilation system will be based off of systems that prairie dogs use in their own burrows.

Letting fresh flow into the structure very easily.

Prairie Dogs live in North America, in underground burrows that they've dug out. Their burrows are very organized with bathrooms, nurseries, sleeping quarters, and even spots for them to hear predators from above! Lots of animals share the Prairie Dog burrows as shelter to live in over night. They live in North America, in underground burrows that they've dug out. Their

burrows are very organized with bathrooms, nurseries, sleeping quarters, and even spots for

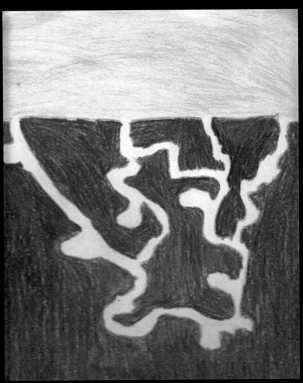

them to hear predators from above! They stay together when they are born, meaning they live in the same burrows, and share food with each other. Prairie Dogs also chase off other Prairie Dogs who want to share their burrow with them. They often greet each other by a kiss, or nuzzle (National Geographic). Prairie Dogs use their mathematical skills to figure out that their burrows need to be so close to the ground because more of a breeze flows on the ground, than up higher on the ground. They realised that the lower they made their burrow the more breeze would come into their home, making it a more suitable environment. The air flows through the lower mound that leads into their burrow, ventilating through holes around 4 centimeters wide, that's smaller than the Prairie Dog itself.

When a cool breeze flows on top of the burrows it enters through the smaller burrow hole, and ventilates through the burrow meaning it goes through all of the different sections of the burrows making it a cool place to live. Once it enters through all of the different sections of the burrows it then flows out of the end of the burrow back outside, and soon enough another breeze will come to do the same exact thing (Bio-mimicron). Prairie Dogs create sharp mounds to make air flow more efficient to come and go inside the burrow, showing how sophisticated Prairie Dogs are with their shape design, and not height because they figured out if they make their mounds a certain way it'll be easier for air to flow in and out of the burrow to give them a cool breeze. Prairie Dog air funnel burrows are around 3 to 4 centimeters, which is very tiny for their size. The tinier the air funnel burrow, the more cool air flows throughout the burrow. Prairie Dogs burrows maintain around 65 degrees fahrenheit during the summer, and around 50 degrees in the winter, showing how intelligent these creatures are, by making such a cool space to live in. Prairie Dogs can make their burrow temperatures just about any degree they want by adjusting the different entrances of air flow in the burrow.

Air Ventilation Layout

The ventilation system will work is by a gigantic air entrances on the bottom of the whole building which the breeze on the floor will enter and funnel throughout the pipes in the building. After it has funneled through the buildings it will then exit the building through pipes leading to the top of the building, letting it have a clear exit to leave the structure. The pipes will be at a 4 centimeter angle, and have sharp edges, letting fresh air flow into the structure very easily and allow it to funnel through different pipes with ease. For one-story houses it will be very easy to funnel through out the house. There will be an opening in the front of the house, the cool breeze air flows into the pipes, and as the pipes are around the house there are openings in them allowing the air to escape, and filter through the house.

All in all, air conditioning is too expensive for today's economy and we need to find a solution to people who suffer through-

All in all air conditioning is too expensive

out the summer with scorching temperatures. I believe with this idea we can make a big change to the world, and save people a ton of money and energy on air conditioning.

Americans spend over 22 billion dollars every year combined on air conditioning

Hagfish Thread

by Chloe Slack

Hagfish, although they are not the most interesting subject and are not talked about much, do have great potential to better our lives. They produce a slime whenever they feel threatened by a predator or even in captivity when humans touch them or try to handle them. Thread is wound up tightly in little

One hagfish thread that is smaller and thinner than a piece of hair, can withstand almost a gigapascal of stress.

knot-looking pockets within the hagfishes slime glands. When the slime is released the thread pockets are too.

The hidden beauty be-

hind this thread is that when it is extracted from the slime it can be wound in small batches into a silk-

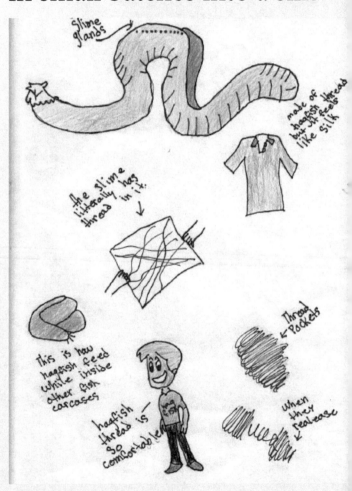

like thread. That thread can then be made into a shirt, or any clothing for that matter. The idea of my product is to minimize rips and tears in clothing, especially when they go through the washer and dryer. Also, if you are active and play sports your clothes won't rip as easily.

Even though it hasn't happened yet, scientists say that hagfish thread has great potential to become the clothing of the future. But hagfish shirts, pants and stockings are probably years away from production. Scientists have already successfully made small batches of cloth with hagfish thread, but it takes a long time to spin it. If we find a more efficient, faster way to spin it, then it would have more of a future in today's world

The reason hagfish thread could make such good clothing material is be-cause the threads in hagfish slime are ten times stronger than nylon. One hagfish thread that is smaller and thinner than a piece of hair can withstand almost a gigapascal of stress. Imagine how many less rips we would have in our clothing if we could find a better more efficient way to make cloth out of hagfish thread.

Remember the next time you buy clothing to look for hagfish thread because pretty soon most clothing could be made out of it. If we could find even stronger materials to make clothing out of that would be really cool.

Shin Ball

by Anna Brown

Did you know that Shark skin is a sandpaper which acts like an external skeleton?

Did you know that french fries kill more people than sharks do and we're not scared of french fries are we? That's why I chose the animal that is a shark. There are many species of sharks, for example, Mako Shark, Great White Shark, and Blue Sharks. Sharks are made for speed and their muscles are like propellers pushing them forward. and that's why I chose them for my prototype. My prototype idea is a sharkskin water polo ball which helps water polo players grasp the balls easier. The reason why the ball gets slippery is because of the water. In the water a ball can get slippery when playing, throwing and catching the ball when it is worn out according to the author Alex Dyer.

The reason why I chose the shark as my inspiration for my prototype is not just because a sharks are awesome, it's because I am a water polo player and sharkskin is great for water polo players. I've been playing water polo for a couple of years now and I wanted to find a solution to improve regular water polo balls. I wanted to improve water polo balls because I was at practice and the balls were only 2 months old and getting slippery and soft even when the ball is dry. I wanted to find a solution to make the balls easier to grip and a ball that doesn't wear out fast. I was thinking of shark skin because it's rough. In scientific research on these elusive predators their skin is in scales and is not like regular sea mammals skin because sharks are cartilaginous fish. Cartilaginous fish means a fish that has a skeleton made of cartilage. Cartilage is flexible connective tissue found in various forms in the larynx and respiratory tract, in structures such as the external ear, and in the articulating surfaces of joints. Sharks are fish not mammals and that is why they have many scales and they don't have blubber.

The way my prototype works is it's a water polo ball and instead of it being covered in leather or rubber fabric it will be covered in shark skin. When you grasp the ball or when catching you will feel a better grip because the shark skin will not slide off your hand. The shark skin

Shin Ball Prototype

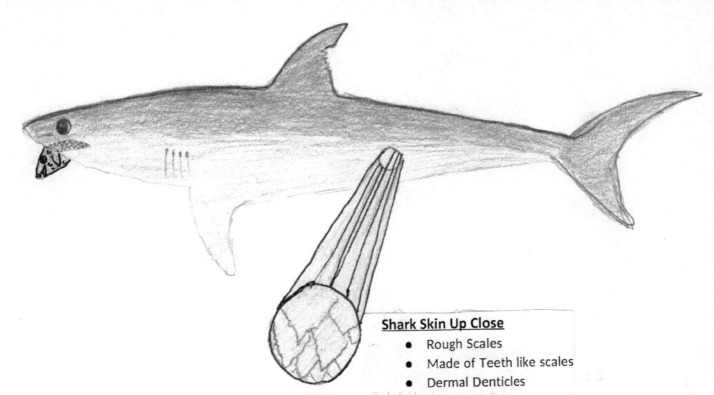

Shark Skin Up Close
- Rough Scales
- Made of Teeth like scales
- Dermal Denticles

is made of "Dermal denticles" which roughly translates to "skin teeth" and rightly so. Their composition closely resembles "mouth teeth" from a shark. Which is also called scales that will give them an improved grip of the Shinball.

This prototype idea is significant because it will benefit a wide variety of water polo players advance in their practice. With regular water polo balls, the wear and tear isn't really good because after a month it starts to wear out. With this sharkskin water polo ball it will never wear out. Using the shark as inspiration for my sharkskin water polo ball is better than the original design because my water polo ball doesn't wear out as fast compared to the original water polo ball.

The reason of why I selected the shark is because of it's unique features. Some of it's unique features are its teeth, eyes, and most importantly it's skin. With the

shinball the water polo players can have a superior restrain on the water polo ball. My design is significant to water polo players because it can enhance their performance.

POLAR BEAR JACKET

By Chester James Ramos

To me, polar bears are very interesting animals. They have many features in their fur which can help them keep warm and cozy.

They have a speacial type of fur that can help them survive years in the cold ice parts of the Antarctic

It is a great way to help those who live in the cold or in winter. My product is a polar bear jacket and this article will show all about polar bear fur and my mimic of fur to make a jacket that can make you warm.

The arctic bear has fur which is like a working solar panel. They absorb heat and the bottom layer of skin, which is black, absorbs energy for them to lose weight and move swiftly. "What is important is the number of hair, and the density. The light will be reflected many times, and some will go back to the skin", says Priscilla Simonis, a physicist at the University of Namur, in Belgium. Their fur was also made for swimming which helps them dry faster than regular clothes or different animal fur. They also have

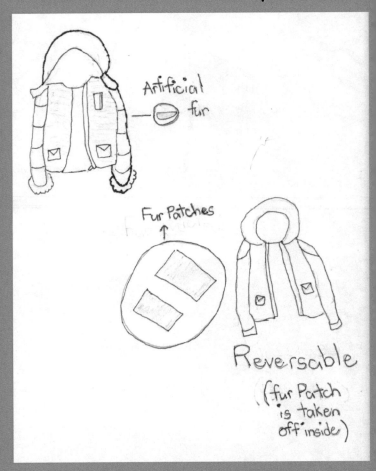

many other adaptations. They are very strong animals and mostly eats seals. Polar bears can run an average of 30 miles per hour and can also swim over 400 miles at a time.

They are the most feared animal in the Antarctic and when standing up can be taller than 10 feet. My product is used to help people who are cold outside. It has two different versions. The regular version is filled with artificial fur that can dry faster and can absorb heat fast enough to keep you warm for long periods of time. For the other version, all you need to do is put the jacket inside-out so it can be a regular sweater, that doesn't absorb as much heat.

This prototype can help people that have outside jobs in the winter. It can also help people who don't want to get wet a lot because you can dry off the water in a couple of minutes. This can be most useful in the winter because that is when the temperatures go down and your body is mostly filled with cold-stress (stress that comes from the cold that can hurt your neck). The way it works is by the artificial fur absorbing heat.

The polar bear is a very useful animal and we can learn a lot from it to help us make life better and easier. My jacket can make people feel warm and cozy in a matter of minutes so they can have a nice day without having to be stressful or cold.

← Nose
(only part without
fur)

The Use of CRISPR

By Kaya Hoffman

Using CRISPR has it's pros and cons as well. If you can make the right cut in DNA, you could cure genetic disease. If you make the wrong cut, you could cause it

Imagine snorkeling off the coast when a teacup-sized colorful blob floats by. You are astonished. Is it a broken off piece of coral? A jellyfish? A sea slug? There are so many possibilities it's astonishing. What you just saw was a nudibranch. Nudibranchs come in many shapes, sizes, and colors. They can look like almost anything! But the most amazing thing about the nudibranch is it's ability to heal. With the use of proteins inside its body, it can regenerate body parts. Scientists are already looking at this to help with loss of flesh. With new technology called CRISPR, this is possible. But will this possibly affect human evolution down the line?

The nudibranch is a surprisingly colorful bottom-dweller that has the wildest tastes in both color and form. These mollusks, part of the sea-slug familia, have the most amazing shapes, hues and patterns of any animal in this world. There are more than 3,000 species of known nudibranch, and new ones are being discovered every day. They can be found in the Earth's oceans, but are more commonly found in shallow tropical waters. Their scientific name, Nudibranchia, means naked gills, describing the feathery gills and horns that they wear on their backs. Typically oval in shape, nudibranchs can be wide or thin, long or short, awesomely colored or bleak to imitate their habitat. They can be as tiny as .25 inches (6 cm) or able to grow up to 1 foot (31cm) long.

nudibranchia
more than 3,000 species
lives in the sea
heals using protiens
Nudibranch (naked gills)

Nudibranchs are carnivores thart slowly exert their range eating algae, sponges, anemones, corals, and even other nudibranchs. To determine their prey, they have two very sensitive tentacles, called rhinophores, found on top of their heads. Nudibranchs get their flabbergasting colors from the food they eat, which helps with camouflage, and some even keep the foul poisons of the animals and plants that they eat, then use them for defense against predators. Nudibranchs are simultaneous hermaphrodites, and can mate with any mature member of their species. Nudibranchs' lifespans vary, some living less than a month, and some living up to a year.

CRISPR-Cas9 is an old technique used by microbes to defend against viruses.

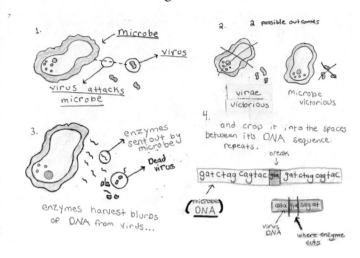

down the line? If we use DNA from animals like the nudibranch, how can we say that the

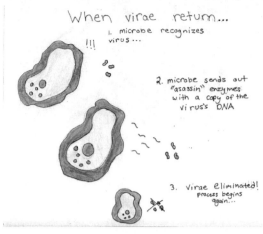

The microbe would typically have to survive a virus attack. Then the microbe would send out enzymes to harvest bits of DNA from the virus' remains. If one of the virus returns, instead of fighting it the same way as the first time, the microbe would send out enzymes with a copy of the virus' DNA. Once the enzymes find the virus, it would eliminate it easily. This was discovered by Jennifer Doudna when researching unusual microbes in an highly acidic abandoned mine. Now it can be used to solve mysteries that we've been searching to solve for years.

Using CRISPR has it's pros and cons as well. For example, if you make the right cut in the DNA, it could cure genetic disease. But if you make the wrong cut, you could cause it. CRISPR also gives us the technology able to modify human embryos. In fact, scientists in China have already done it using embryos that cannot be born, even though they can go through the first stages of development. But the biggest problem with this is ethics: Do we really want to modify the future's children without their consent? I mean, they haven't even been born yet. And the biggest problem of all: How will this affect human evolution

future effect oo0[]
\f this will not be nudibranch-people walking around?

As of right now, we can only observe where this raging battle about the ethics and future of CRISPR will head. Who knows, maybe in the distant future we will have nudibranch-people and genetically modified babies. CRISPR might be the end of cancer, or alzheimers, or even autism. We can only observe where CRISPR will take us. We can only observe…

Tuna Drone

By: Sam Aguirre

Did you know that just one quart of oil can contaminate 250,000 gallons of water?

Oil spills are a huge problem and it kills millions of marine life. It also can affect us humans. Something is desperately needed to fix this. I came up with an idea of having little drones about the size of bluefin tuna. They can cruise around areas of the ocean with oil spills to filter the dirty water in the ocean.

I am modeling my invention after a bluefin tuna. The bluefin tuna is a very interesting fish. One adaptation that the bluefin tuna has is that they have opposite coloration which lets them be able to blend in with the water surrounding them. This adaptation makes it harder for predators to find them. They are dark blue on the dorsal side of them which lets them blend in with the dark ocean below when seen from above the tuna. Their ventral sides are white which allows them to blend in with the bright sky when seen from below.

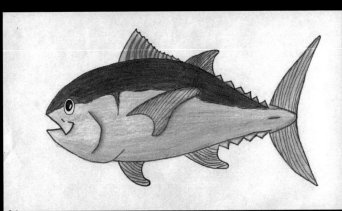

A couple more adaptations the blue finned tuna have is that they have aerodynamic bodies. They also have flushed eyes and head with retractable pectoral fins that help lower the amount of drag when they are swimming. They use their fins like the wings of a plane to generate lift. These fins are also used to steer when they pursue prey. Thier lunate-shaped tail is very helpful. Also the rows of finlets on bottom and top of the posterior section of their tail which help lower the amount of drag. Tuna have very small cycloid scales which lets them have smooth skin. They also have streamlined fins. This lets them be able to make really quick movements.

As I previously mentioned, my invention is able to filter water. I got this idea from another invention called the lifestraw. The lifestraw is a straw that is able to filter almost any liquid into clean water safe enough to drink. In the lifestraw, there are 4 filtration systems in the straw. The first one is a textile pre-filter, the tiny openings in the mesh of the filter mesure 100 microns (which is roughly as thick as a single strand of hair) , this is able to filter bigger particles, like sediment and dirt. Next, the water then passes through a polyester filter. This is like a mini sifter. This is able to filter clusters of bacteria as reported. From there, the water transfers through a chamber of beads that are saturated with iodine (A chemical element of atomic number 53, a nonmetallic element). The iodine is able to kill 99.3% of all bacteria and viruses. And last, the water goes through a chamber of granulated active carbon. This will kill any remaining bacteria there is left. The problem is that the lifestraw leaves behind the bacteria. What I need is to find a way where it will leave behind the water and store the bacteria. I am also using adaptations that tuna have like it's opposite coloration and the fins and scales they have to help reduce drag. Instead of one lifestraw like system, I can have four with the water entering the drone similar to the tunas gills to get more filtering done. The filtered water then exits from the tail. Colonies of these drones could be cruising around oceans.

My invention is important because the pollution in the ocean kills millions of marine life and can affect us too The oil can get into the fish that we eat. About 80% of the pollution in marine areas comes from land. One of

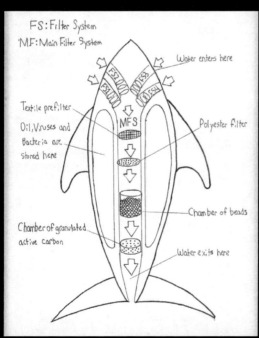

the biggest sources is referred to as nonpoint source pollution which results from runoff. The pollution of Nonpoint comes from little things like, trucks, septic tanks, boats, cars. It also comes from large sources such as ranches, forest areas and farms. The engines from

motor vehicles drop droplets of oil each day on roads and parking lots which eventually makes it's way to the sea traveling through rain runoff. Air pollution can also cause water pollution, which can settle into waterways and oceans. Also dirt can pollute oceans. Silt or topsoil from construction sites of fields can run off into waterways harming wildlife habitats and fish. Humans and wildlife can be affected by nonpoint pollution of the pollution in oceans and rivers.In some cases, pollution can get so bad that areas can be closed after a rain storm. In the U.S, coastal pollution affects more than one-third of shellfish growing waters. There is evidence that ocean pollution has dated as far back as Roman times. Some common pollutants that are man-made that reach ocean are chemical fertilizers, oil, pesticides, detergents, herbicides, plastic, sewage and other solids. Lots of these pollutants gather at the bottom of the ocean, where marine life eat them and are later presented into the global food chain.

I could have chosen any kind of fish to use as inspiration but tuna was the first thing that came in my mind mostly because of it's size. The tuna has adaptations that really unique and interesting. My invention is important because pollution is a terrible thing. If you saw how much pollution is in the ocean, you wouldn't want to go in the sea. It also is really sad that we kill all sorts of animals because of us. But with my invention, we help fix that little more. The ocean would be a little cleaner and a little better.

G.I.P.C

(Giraffe Inspired Produce Cart)

by Amanda Garrington

Did you know that over 6,000 people die from falling off ladders each year and more than 30,000 people are injured? Sydney Newbert wrote an article about this subject on prezi. This is a big problem for farms and fruit organizations, especially on uneven ground such as dirt and rocks. What would you say if I told you that the solution is a giraffe?

Picking apples from a normal size tree can be very intricate and a ladder most likely will be involved. A dwarf apple tree gives 48 pounds of apples (a bushel) a year while a natural apple tree makes about 384 pounds (8 bushels) of apples each year. Which one do you want if you are selling them? These companies only have dwarf trees so that they can have a "Pick Your Own" program. But just think if we had something so that they could pick from the big trees without it being dangerous, they would save more land for more apples and no one would get hurt.

Over 6000 people die a year from falling off ladders, and the solution could be a giraffe.

There could be a solution. I used a giraffe as the inspiration for a product that can take away the risk of death or injury caused by using a ladder, and saves more room and more time. Giraffes have a prehensile. Prehensile means that it is able to grab or carry something. A giraffe's tongue is about 50 centimeters to 20 inch and there upper lip is also prehensile which makes pulling leaves, fruit and branches very easy for them. A giraffe eats hundreds of pounds of food a week and have very thick tongues. They also have very thick spit so the food sticks to it. We could use all of these adaptations to create a solution to the fruit picking problem.

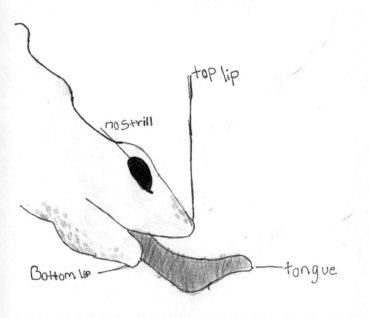

"G.I.P.C" (giraffe inspired produce cart) is a cart with 2 robotic arms made of Tungsten metal and other various types of metal. These robotic arms have tips modeled after giraffe lips and mouth

for a safe grip and productivity. This product can help be efficient, reduce life loss, and save room so we get more produce in less time. This also means that companies can get more money and be more successful. For their "Pick Your Own" program they can have certain carts dedicated to the program.

So as you can see, this machine

can benefit people in many ways and prevent unexpected injuries that could be fatal, especially when it comes to produce. This product can solve three different problems and could make a huge difference when it comes to saving lives.

Titanium Sub

by Kendra Fuller

Do you know how sea otters swim and how slender and flexible their bodys are? Also, did you know that river otters range in size from about 87 to 153 centimeters? Because they are small they can fit through small places. Also, did you know that sea otters have similar shaped bodies in comparison to other otter species sea otters are somewhat with larger ribcage.

The reason why I chose this animal is because of the the way that it swims, the way it is shaped and the way it goes side to side when it swims. They have really slender and flexible bodies. According to Ria Tan from Naturia, "they can swim for long distances and stay underwater for 6-8 minutes with a single breath". Otters use their strong, long tails to push them through the water when they are swimming fast. They paddle with their powerful webbed front and back paws

when they are moving through things underwater. Ria Tan also says that "otters are also excellent divers" (Naturia). Their whiskers can sense movement in the water

which helps them to find their prey when they can't see in dirty or dark water.

The prototype is going to shaped like a submarine but it is going to have a tail and robotic arms to get samples of the ocean and maybe even find new life forms. It is going to be able to get bigger and smaller. Since sea otters a very flexible and narrow they can fit through small spaces, so I want my submarine to be able to allow the people inside into every cave or small place in the ocean. It is going to swim like a sea otter. It is going to be made of a spherical titanium hull. I picked this because it is pressure resistant and lightweight. In the middle of the submarine is going to be chamber where the people will be in order for the submarine to get bigger and smaller only around the chamber will get smaller so that the people will not get crashed.

The innovative submarine is important because when you are in a submarine you can miss a lot of things in a big metal contraption .

That's why I wanted to make a submarine that gets smaller and bigger so you can go through small caves. The tail of the sea otter is a huge part of the way it swims, so adding it to the prototype will allow it be able to swim around in a circle like a sea otter. Also, the robotic arms are based on the arms of the sea otter it is used on the submarine so that it can get samples of the habitat.

In conclusion I feel this is the best way to be able to see the whole ocean and really see and feel the real experience of the sea. Also, this submarine can get so much more than just what is outside of the caves and small places. Sometimes animals like to hide so that's why you can't always see them. This submarine will really help find them.

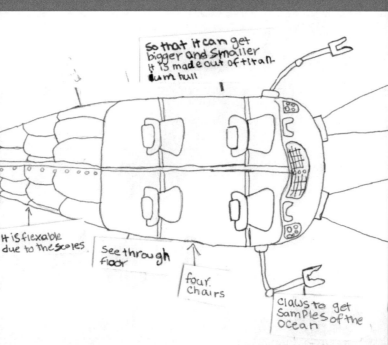

So that it can get bigger and Smaller it is made out of titanium hull

It is flexable due to the scales.

See through floor

four chairs

claws to get samples of the ocean

The Sharkboard

By Jackie Boyce

Leopard sharks are fast-swimming, strong, vicious, and tough creatures. This animal is one of my main sources of inspiration for this project. My prototype, a surfboard with a leopard shark skin-inspired outer layer, is important because it prevents injuries. The surfboard is more stable, balanced and can float in the water better because of the leopard sharks shape and design.

"The surfboard will be made out of polystyrene foam...the result is a light and strong surfboard that is buoyant and maneuverable."

Leopard sharks skin is made of thousands of small, hard, structures shaped like razor sharp teeth that make the skin rough and very hard to cut into. My prototype uses leopard shark

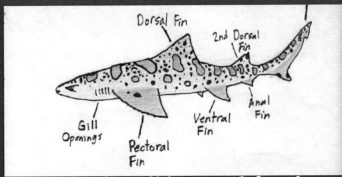

skin which will be used for the grip. The shark's pectoral, anal and caudal fins (the fins on the bottom of the shark) help it remain stable while swimming only inches above the sea floor. According to Shark Sider, "Leopard sharks generally grow up to 4-5 feet long". They have long, narrow bodies so they easily glide through the water. "They can go more than 8 mph but tend to cruise around 1-4 mph," as said by Demand Media.

My prototype will work just like any surfboard does but it will lead to people being safer, more balanced and stable on the surfboard. The surfboard will be made out of polystyrene foam (Polystyrene is a synthetic aromatic polymer made from the monomer styrene) with layers of fiberglass cloth on the surface, the result is a light and strong surfboard that is buoyant and maneuverable. The leopard shark's fins will be what the surfboard's fins are based off of. Their improved design will be much easier to ride and stand up on. It will be easier, and I think that it will inspire people to get more exercise, build another hobby, and have fun.

Leash

Grip

Gripping Pad

It's narrower than most boards which makes it easier to go through water

· People get knocked unconcusions, drown and get seriously injured by falling off of the surfboard because it's unstable so when they stand up they fall

Grip/gripping wax

Birds Eye View

· Thousands of people die each year from their injuries from surfing. It's not exactly surfing that is dangerous it's mostly how good the rider is and how updated the board is.

other ways of people getting hurt that it won't be able to prevent. We can make it so less people will fall off, get a concussion, drown or get seriously injured. According to ask.com, at least 21 people a year die from surfing injuries. I think this animal is the right inspiration because of how many ways we can use it to improve the original design and tie nature into it.

This animal has so many great adaptations it has created over the years that can make this invention even better than it already is. I think this updated invention can prevent injuries, make surfing easier and much more. I want this board to be a new way of being safer, enjoying yourself, having fun and connecting to nature.

We need this prototype because of all the injuries that keep happening. There will always be

Booca

By: Blanca Salas

Did you know that humans consume about 27,000 trees daily just for toilet paper? I have designed a new type of paper to help save trees. After doing some research, I've found out that Bamboo and Sugarcane are the two fastest growing plants. My design could save about 34,000 trees per day.

Did you know that Bamboo and Sugarcane are grasses, not hard-

woods? Trees are hardwood. These two grasses grow very quickly and after harvest, also known as growing crops, they grow back really quickly. According to Stéphane Schröder, the author of Guadua Bamboo, "Of all grasses, bamboo is the largest and the only one that can diversify into

forest". For these plants, replanting is not necessary. Unlike trees. Trees take up to 30 years to re-grow and mature, when it takes both bamboo and sugarcanes 1-2 years to reharvest. According to the inventors of Caboo, creators of toilet paper and tissues made of bamboo and sugarcane, "Bamboo's natural rhizome (root) network protects soil from erosion, wearing away sand, soil, or rocks by water or wind, and retains moisture, and bamboo can grow in environments with depleted soil and little water. In fact, it actually returns nutrients to the soil improving degraded areas". Not only do Bamboo grow faster than trees, but they also provide 35% more oxygen than trees.

First, the grass has to be cut. Second, The wood is cut into small wood chips. Third, small pieces of wood are mixed with water and are made into a mixture. Fourth, the mixture is bleached and chemicals are added to make the paper stronger. Fifth, the mixture is flattened and the water is squeezed out. Sixth, the mixture that is flattened is then dried, and becomes paper. Seventh, the paper is cut into smaller pieces. Where can you find Bamboo and Sugarcane? According to Stéphane Schröder, the author of Guadua Bamboo, "Bamboos grow in the tropical and sub-tropical regions of Asia, Africa and Latin America, extending as far north as the southern United States or central China, and as far south as Patagonia. They also grow in northern Australia." According to Sugar Industry Biotech Council "The main sugar cane producers around the world are Brazil, India, China, Mexico, Australia, Thailand, Pakistan

and the United States. In the United

Did you know that humans consume about 27,000 trees daily just for toilet paper?

States, sugar cane is grown in Florida, Hawaii, Louisiana and Texas."

My prototype is very important because it will save many trees. We need to save trees because sooner or later we are not going to have any more trees. Because we use so much paper we may not have trees in the future. Humans consume about 27,000 trees daily just for toilet paper. That's like wrapping paper around the world 118 TIMES EVERY DAY! Why is my paper better? Let me see…. My paper is better because it will save over 34,000 trees per day.
 It's also better to use my type of paper because it's more eco-friendly. This paper is even 100% biodegradable (decomposes naturally).

I chose to use these plants because they are the two fastest growing grasses. I also got inspired by trees because we are using too many trees and we need to put a stop to that. My design can help humanity become more environmentally friendly while saving money.

360 Vision

by Quentin Callahan

The 360 Vision glasses allow the user to see behind them. This prevents sneak attacks, which would be a threat in military situations. This can improve life by keeping it's user safe. If soldiers can see behind them, they can avoid being attacked.

The Woodcock is the name of a couple different species of birds that are capable of seeing 360 degrees. The Woodcock's eyes are located on the sides of its head. This gives it all around vision, but it's only capable of binocular vision (good depth perception) directly in front and directly behind it. This adaptation is used to help the Woodcock remain elusive from its predators.

The 360 Vision glasses are similar to normal glasses. On the end of each arm of the glasses, that hook around the back of the user's ears, is a small camera that feeds video to one lense on the front of the glasses. Each camera's' video is stitched together, otherwise the image would

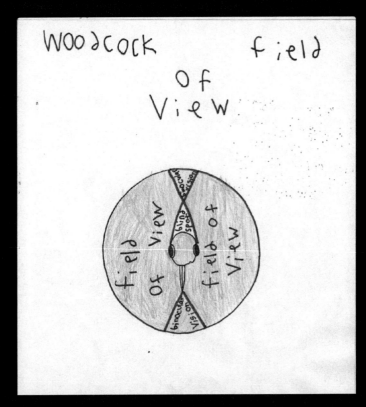

seem incomplete. Humans are not used to 360 degree vision, so the background feed is only in one eye. This allows the user to see both forward and backwards simultaneously.

My 360 vision goggles will allows the user to see behind them, preventing most sneak attacks.

It is not uncommon for a soldier to be attacked from behind in battle. Humans can't see behind them, so this is a huge threat. My 360 vision goggles will allows the user to see behind them, preventing most sneak attacks. I chose to use the woodcock because it can see behind it. I chose to make this so soldiers don't get attacked from behind.

Why do we need this? We need this device because I feel that it would cut down on our soldier's death rates. In conclusion, this device can help soldiers in saving many people as well as prevent soldiers from being attacked from behind.

Suction Gloves

by Atticus Dixon

Did you know that an octopus can grow up to 15 feet long but can maintain only 100 pounds. They also have a short lifespan lasting between 6 months to 5 years. My prototype is a pair of suction cup gloves. This can improve the lives of firefighters to climb up the wall instead of running through the burning house to get upstairs or even something as simple as helping window cleaners climb taller buildings.

The animal that inspired this design is an octopus. I used this animal because it has a good resemblance with my idea, the suction cup gloves. I learned a lot about the suction cups on the octopus and what they are used for. They can be used for staying close to the bottom of the ocean, sticking onto reefs, and they can even sense and taste food. Obviously my gloves can't taste food, but they have similar uses to what an octopus uses them for.

The way my prototype works is simple. They are just a pair of gloves with suction cups. All you need to do is put the gloves on and use them on a window or some sort of smooth surface.

"The suction cups can be used for staying close to the bottom of the ocean, sticking onto reefs, and they can even sense and taste food."

The glove will be made of rubber and the suction cups are made of PVC plastic. Rubber for the glove will be more prefered to some sort of fabric because they will be more stable on your hand and PVC plastic prefered to rubber because it is a stronger material for climbing for in my case, it would be climbing.

These gloves will come in handy when firefighters have to

get through a two story house to get people out of the house or they could be used for window cleaning for getting higher spots that can help in cases of fires in large buildings or even something like window cleaning. I'm hoping I can get this idea a little

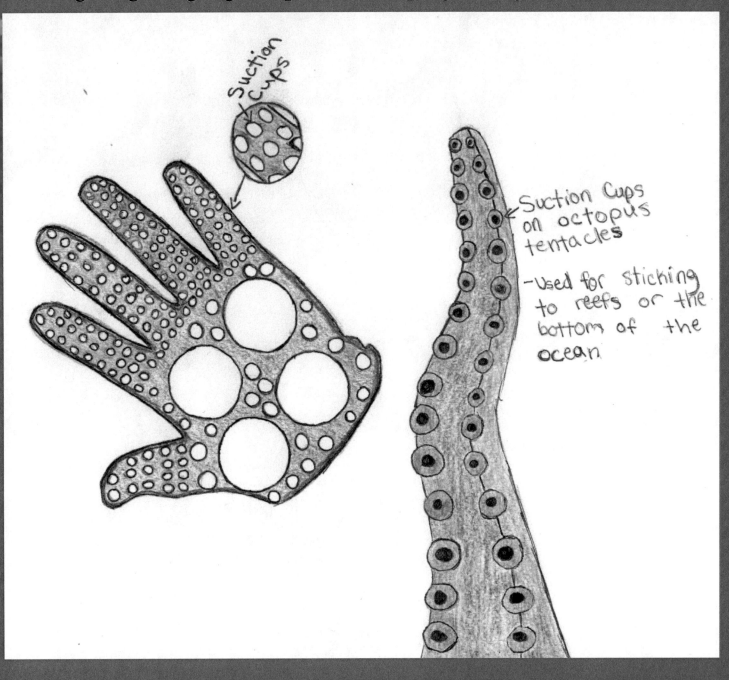

Suction Cups

Suction Cups on octopus tentacles

-Used for sticking to reefs or the bottom of the ocean

on the window or to clean taller windows to where you can't reach.

This animal was a great inspiration for suction cup gloves bit further out to a company so this idea can hopefully become my new invention.

Lizard Shoe

by Katrina M Estrada

What would you think if you could be able to walk on water with water shoes? Well, my product allows you to walk on water as you please and keep you balanced. My product is useful when creating bridges,and when you are stranded in the middle of a lake you can just walk back to shore. This could be useful because sometimes boats can flip over and you might not be able to swim, then you drown. My product is a life jacket with a tail and shoes that will keep you afloat and alive.

My inspiration was the Green Basilisk Lizard which has the ability to run on water. The bottom of their feet are surrounded (fringed) with scales that spread out when it hits the water so it does not break the surface film of the water. Also, they can run up to 25 mph and they move their legs like a windmill. They have long tails to balance and help when running and there

tails move side to side to help support both sides.

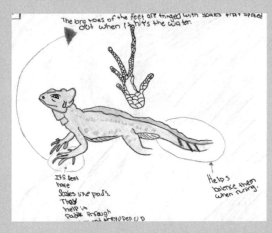

My main product is the shoes which have an air compressor in the sole.The scales will be made of mini floatation devices that holds one extra pound of buoyancy. Buoyancy is the upward force that keeps you afloat. The boots will be made of water resistant rubber, the life jacket will be made of water resistant plastic with air. The tail will be made of water resistant metal and the it will swish side to side because of a fin that will be at the end of the tail.

My shoes and tail as well as the life jacket can save a life and

These will be for walking and safty.

The scales will see thade of your floaties.

There are 6 scales because each holds one extra pound of buoyancy and if you count both boots that's 12 extra pounds of buoyancy.

Looks like normal boot from side and front but..

Side View

Scales in inner insole

Scales in sole

bottom of the shoe

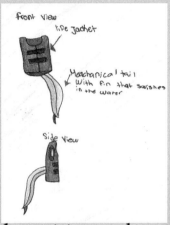

front view

life jacket

Mechanical tail with fin that swishes in the water

Side View

a tragedy waiting to happen. This story is written by Tricia Manalansan, "Liam Vaughan a three year old boy from Rockford, Illinois was visiting a relative in the city of Morris on Sunday. The house did have an above ground pool in the backyard that the family used. Vaughn wore inflatable floaties for most of the time he was in the pool, but the boy took them off when he went to eat. Vaughn returned to the pool disregarding that he needed to retrieve the floaties. At that same moment, his mother left to check on the other children inside of the house and returned outside mo-

ments later. Her and her husband noticed the toddler was nowhere to be found in the backyard until they saw him motionless in the pool. Liam Vaughan was pronounced dead from the drowning at 4:11 p.m. at the emergency room at Morris Hospital." This shows us to always be careful and to put safety first. My shoes and tail both have floaties and can save your life, even if you can swim when a boat flips over, you slip when fishing, you can hit your head when you fall in the water, or when you are in the water. This is so serious and even if you're a professional, things happen and you can't stop them, so that is why my tail is attached to a life jacket and my shoes have buoyant scales.

In conclusion, the Green Basilisk lizard was so incredible, that is why I chose it. It has so many cool features and I think my product can save a life from an incident that can turn into a tragedy.

What would you think if you could be able to walk on water with water shoes?

Slide Off

by Jonathan Cardona

According to Kaady Car Washes, "if you wash your car at home you can be wasting 140 gallons of water each time you wash." Imagine if you didn't have to waste all of that water just to keep your car clean. By using elements based on the qualities of the Pitcher plant, an insectivorous plant that has special leaves that help it trap insects, and using it to coat the outside of a car, we can prevent it from getting dirty. So we can save water by not washing the car.

When wet, the Pitcher plant's surfaces are slick and insects aren't able to come out. A pitcher plant makes insects fall when they stand on the inside of its leaf. Insects fall inside of the Pitcher plant because of two reasons, its funnel-like shape, and the slippery substance covering the leaves (Sciencedaily.com).

The adhesive element in all car paints is the part that makes them stable and permanent. Combining this adhesive element with the slippery substance from

"By using elements based on the qualities of the Pitcher plant...we can prevent it from getting dirty. "

the pitcher plant would make car paint durable and resistant to bugs, dirt, and rain. In addition to painting cars, we could use this substance to coat the windows of big buildings. This would help buildings have clean windows and make it so that they wouldn't need window washers. In addition to coating windows of big buildings, we could use this texture to cover furniture. This would help people have clean furniture if drinks or anything else

spills on it. They wouldn't have to worry about stains.

California is facing one of the most terrible droughts on the record. Governor Brown announced a drought State of emergency (CA.gov/drought). If necessary, change your behavior when it comes to using water so you can be prepared for the water shortages. This design, based

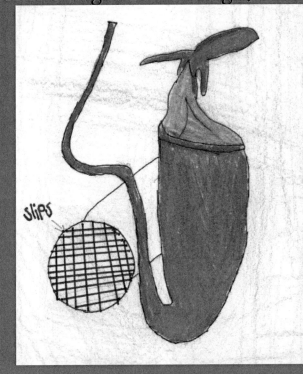

slips

off of the Pitcher plant, allows you to coat almost any hard surface with the slippery substance. This slippery substance can help prevent the waste of water because it keeps most things free of dirt.

By using elements from the Pitcher pitcher plant to coat any hard surface, we can prevent those surfaces from getting dirty. When these surfaces don't need to be cleaned regularly, we will save water. Overall if families used this design on their cars, they could save on average 365 gallons every year.

Limb Regeneration
by Philip Alioto

Imagine all the veterans that have lost body parts to protect our country, the babies born without an arm or leg, and the victims of horrible accidents and diseases that cause amputation. Now, imagine them with fully functional body parts. This dream will be accomplishable with my product. My product mimics the way the axolotl regenerates limb(s) so perfectly to help amputees. This will let amputees all over the world not worry about the struggle of being an amputee and live a happy life where everybody has every body part they need.

I am using the same way that Mexican Axolotls can regenerate almost all body parts . No matter what body part you cut off or destroy, it is able to regenerate. The Mexican Axolotl has the best regenerative powers out all the newts and salamanders. In one lab they conducted organ transplants from one axolotl to another from different families and they had no problems accepting one another's organs. Axolotls are found in only found in one place and that is a lake in Mexico that is called Lake Xochimilco. Because they are only found in that one place on Earth, they are very endangered and not many labs do experiments with them. James Goodwin of the Australian Regenerative Medicine Institute says. "We need to know how salamanders regenerate and how they do it so well to reverse engineer it into human therapies."

Amputation is the surgical removal of all or part of a arm, leg, fingers, hands, or toes. 1.8 million Americans are living with the struggle of being an amputee. The most common form of amputation is either above or below the knee. In many cases amputation must be necessary. The most common reason of amputation is poor circulation because of damaged or narrowing arteries. This disease is called Peripheral Arterial Disease. This causes inadequate blood flow causing the body's cells to get insufficient amounts of oxygen and nutrients they need from the blood stream. As a result to this, the affected tissue begins to die and might cause infection to spread. Other causes of amputation of limbs is suffering from a severe vehicle accident or serious burns. Or you might need an amputation because of a cancer ous tumor in your bone or muscle of your limb. A serious infection in your tissue that is resistant to all antibiotics and treatments called neuroma thickens the nerves tissue and cause amputation. The last example of how you can get a amputation is being in severe cold and get frostbite.

"We need to know how salamanders regenerate and how they do it so well to reverse engineer it into human therapies."

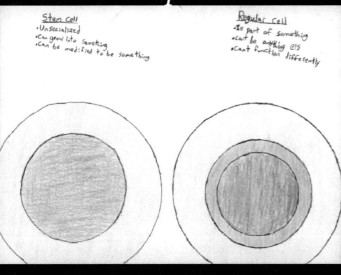

The way this regenerative medicine works is that you are are given a surgery if you had to amputate a limb. This is so the surgeons can reconstruct your nerves and veins to fit where they normally go to have proper reflexes and blood flow. If you were born without a limb or limbs then this may not be necessary to have the proce-

The reason I am choosing to use the axolotl regeneration as inspiration is because they can regenerate almost every part flawlessly. Axolotls are the best when it comes to regenerative power. Not only are they the best when it comes to regeneration, They are also 1000 times more resistant to cancer and other diseases than any other newt or salamander. Because they can regenerate limbs very quickly and perfectly, and be much more resistant to disease, this makes them the most efficient to mimic out of all the newts and salamanders.

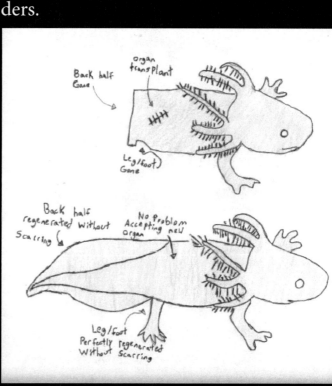

Snake Ointment

By Hector Rosas

"Did you know that snakes will get rid of their old skin and get new clean skin?"

Did you know that snakes will get rid of their old skin and get new clean skin? Imagine if we could just put some ointment or cream on our face, wait a few hours, and just start peeling our old skin off! Wouldn't that be cool? By using the idea of shedding snake skin, we could allow people to shed their skin.

Snakes are reptiles that do not have legs and will shed their skin a few times a year to get mites, dirt and other things off their skin resulting in new skin. This process will happen their whole life. They can live from 20 to 50 years and do not have eyelids so you can't really tell if the snake is staring at you or sleeping.

Snakes shed their skin every once in awhile so their skin is nice and not really dirty. We can take the snake's technique of shedding skin and create an ointment that does the same thing to human skin. In that case we could make a cream that we could apply as a mask. It is even possible to get that part of snake DNA and remove the part that we don't need like, in Spiderman a guy put lizard DNA in himself and grew his arm back, but then turned into a giant lizard. In this case maybe our skin would shed and then have nice skin. It would be amazing if we could shed our skin and just get it off and have new clean smooth skin.

It would be helpful for some

people with skin problems or some abnormal things in their skin. For some it maybe acne or a slight skin disease that may be dangerous. This product would be better than some products that have lots of chemicals and can leave you even worse than before you used that product.

I chose the snake because I thought the way they shed is cool, and I feel that we need this because it's something that some people want and would help them. Also, this could cure some light skin disease and would be made naturally not with those danger-ous chemicals that can do harm to your skin. Surgeries cost a lot so it would be at a non expensive price.

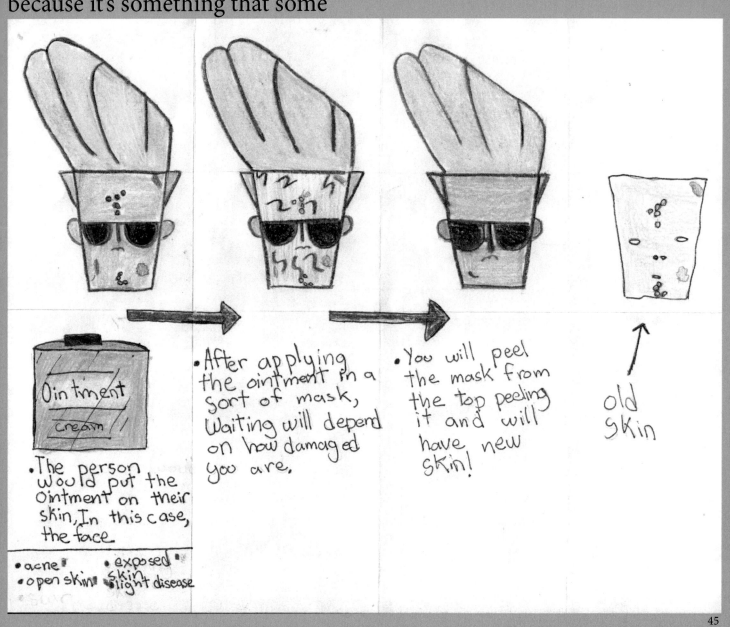

Ointment
cream

• The person would put the ointment on their skin, In this case, the face.

• acne   • exposed
• open skin   skin
          • light disease

• After applying the ointment in a sort of mask, Waiting will depend on how damaged you are.

• You will peel the mask from the top peeling it and will have new skin!

old skin

Naturally Clean

by Maya Tamir

Cleaning water is expensive and takes a long time, but what if we could make it more simple? Home Advisor states that on average, people have spent over $1,500 on cleaning water in 2015. Right now, huge machines clean our water, but that makes it expensive and creates pollution. Some companies are already starting to use phytoremediation, cleaning water through plants, in their water filters because of how well it works in marshes and wetlands. Marshes are areas that are really wet and dense with vegetation, feeding clean water into a pond or lake nearby.

> *People have spent over $1,500 on cleaning water in 2015.*

Marshes have started becoming the inspiration for many sewage treatment programs because its methods of water filtration are so effective. The National Park Service states in an article on water filtering of wetlands, "In wetlands, water flow slows, so suspended sediment drops out and settles

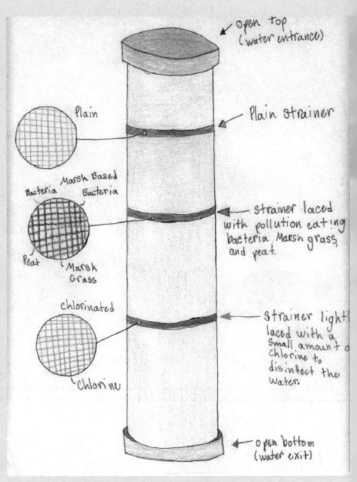

to the wetland floor." Roots from the plants will also pick up some harmful chemicals in the water, and make them less toxic, or turn them into energy. The National Park Service states, "As water flows through a salt marsh, marsh grasses and peat [a spongy matrix of live roots, decomposing organic material, and soil] filter pollutants such as herbicides, pesticides, and heavy metals out of the water, as well as excess sediments and nutrients." If the lake or pond the marsh feeds into has too many nutrients, the wildlife living there can be harmed.

A marsh-based filter will have two strainers, one that is plain, and another that uses peat, roots, and genetically altered bacteria that replicates grasses, decontaminators, and more roots. The second strainer will be located below the first one. The National Park Service states in an article, "A growing industry of bioremediation is being developed from these wetland bacteria which 'eat' pollution."

Roots from the plants will also pick up some harmful chemicals in the water, and make them less toxic, or turn them into energy.

When water is pumped into the entrance of the filter, the first strainer is plain, and will filter out the basic dirt and particles. Next, the water will go through the second strainer that is laced with roots, marsh grasses, peat, and altered bacteria which will 'eat' the tougher chemicals and metals. Lastly, the water will pass through the final strainer, which is lightly coated with chlorine, to ensure that the water is completely clean and free of infectants.

This product doesn't have a designated size and can be built in proportion to the amount of water that is being filtered. For example, the filter in a water tower would be significantly larger than the one inside of a 16 ounce water bottle.

This filter could be very helpful in places where there is drought, or is far away from any water treatment facility. Water.org states that, "Twice the population of the United States lives without access to safe water." If we could create cheaper, more effective, and safe filters, it would be easier for people who otherwise wouldn't have water to get access to clean water.

As a recap, a filter based on the way marshes use plants, bacteria, roots, and peat to filter water would be a more effective and cheap way to filter water and get it to those who need it.

Claw Gloves

by Patricia Martinez

Do you like mountain climbing? Well, I've come up with a prototype that will be helpful for you to use when you're mountain climbing. This prototype is inspired by mountain lions. Mountain lions are really good climbers and they have sharp claws so that's how I got the idea to design gloves from

Their rear legs are the biggest (proportionately) in the feline family and their paws are compellingly enormous.

\their claws. This is important because it's helpful and it can prevent climbers from slipping or misplacing their hand.

Mountain lions, otherwise called cougars, jaguars, or panthers, are profoundly proficient predators. They are talented climbers, most of the time staying in trees. (My prototype can also be used for climbing trees or even ice.) Mountain lions like to stalk from above, utiliz-

Mountain Lion

They are really good climbers and their abilities give them bouncing and sprinting capacities.

- Claw gloves inspired by Mountain Lion paw claws.
- The element of this animal that add to my design are just the mountain lion's sharp claws.

ing rock edges and above landscape. Their rear legs are the biggest (proportionately) in the feline family and their paws are compellingly enormous. These attributes give them noteworthy bouncing and sprinting capacities.

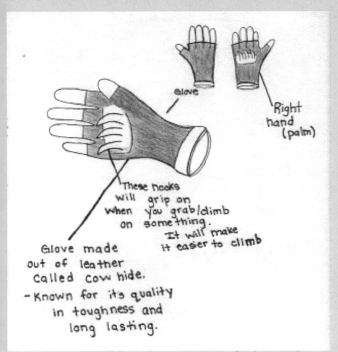

These hooks will grip on when you grab/climb on something.
It will make it easier to climb

glove

Right hand (palm)

Glove made out of leather called cow hide.
- Known for it's quality in toughness and long lasting.

The design of the claw gloves have three claw shapes as a animal claw on the palm of the glove. It will hook to the surface of what it's grabbing. The claws are made out of strong metal. The gloves are made out of a leather material called cowhide. The most regular use for cowhide is as a part of work gloves.

It is known for its quality and toughness. It keeps up its honesty and takes the state of the wearer. As you can tell, the features of the animal that contribute to my prototype is of course only the claws of the animal. The mountain lion species is known for their sharp claws.

These claw gloves are important because it can prevent climbers from slipping. We need this because it can be helpful and protective when you're mountain climbing. Using an animal as inspiration is better than a regular design because you can base off your design on something in nature.

The animal is a good inspiration for my design because it involves with animals (nature) and it easier to base it off of the animal's features and abilities. I require this because it can be helpful and safe when you're mountain climbing. Utilizing a creature for motivation is superior to a general outline on the grounds that you can construct off your design of something in nature which is biomimicry.

Crab Claws

by Anna Bacal Peterson

Have you ever had trouble opening your pickle jar? Almost everybody has been in that annoying situation where you can't open a jar of food or other things. It's pretty frustrating. Well now here's an easy solution. My prototype is

Crabs claws have a material on the tip that helps them have strong grips.

a jar opener based off of crab claws.

Male crabs have one big claw and one regular sized claw while females have two regular sized claws. The crabs have a different material on

the tip of their claws that help them grab and pinch. If they did not have that, they wouldn't be able to crab or

pinch because the claws would get dull.

The crab's claws and legs are made of a strong bromine-rich material that can be used to make different products. Crabs claws have strong grips because of the strong material on the tip of their claws.

When hot stuff is put in a ar the metal lid will expand. When it cools, the lid contracts. If you run hot water on it, it will expand. In San Diego we are in a drought and water takes a long time to heat up, so my invention would be useful.

My prototype will be a adjustable crab claw attached to a handle. You can have the

Almost everybody has been in that annoying situation where you can't open a jar of food.

claw hook on the lid any way that is convenient to you. If your jar has a twist lid, you can twist the handle around

the lid and the cap will come off. If your jar has a pull lid, you can pull down on the handle and the cap with pop off. The claw is adjustable so it can work on all jars.

My invention will help people because it's really annoying and frustrating when you can't get a jar open and most people like to do stuff successfully. Crab claws are sharp and have very strong grips so it would be better to use this than most jar openers.

Crabs claws have a material on the tip that helps them have strong grips. If you connect that to the and twist the cap pops right off. So no more pesky jars. Now all jars are easy to open.

UV Spray

by Ismael Gil

Are you always losing your things around the house? Then UV Spray is for you. My brand new product could help you out a lot with finding lost items. My idea is simple, the spray comes in a can. Spray it on anything that is nonliving, and now that item that you have sprayed will glow in ultraviolet light. This can does come with a ultraviolet flashlight, so no need to buy one. My inspiration for this product is the mighty scorpion. When under the rays of ultraviolet light, the scorpion starts to glow. Unfortunately, the scorpion does not glow when newly molted, or when it has shedded old skin, but the glow does return once the skin of the scorpion has hardened.

•UV stands for Ultra violet.

The way that this spray works is that, we have made a very thin layer of Mr, Clean soap, and liquid cotton. Those are two products that will glow under ultraviolet light. As you can see in the illustrations, you can shine the UV light on a scorpion and it will glow. If it does not glow, then that means it has shed its skin not to long ago.

So, I think my design is a good

Are you always losing your things around the house? Then UV Spray is for you!

idea because it helps a lot of people who are always losing things or who are worried about losing things that are important to them. The reason I used an animal for my inspiration is because nature has already perfected every single little thing that it has to offer, and we humans have not perfected everything. So that is why we

need to take inspiration from nature to be successful.

The reason I chose the scorpion as my inspiration was because they glow under UV light so I wanted to mix other products that glow to make it glow even more. That makes it even easier to see under blacklight.

In conclusion, the scorpion will glow under ultraviolet light, the spray could help you find lost things, It could have potential to to do more than just help find lost items.

surface of scorpion glows under Blacklight

so I used other products that glow under black light

Streamlined Shaped Boat

By McKealy Hayes

> *My prototype is very important to the problem because animals each day die because of boat propellers.*

Imagine you are Green Sea Turtle swimming through the Atlantic Ocean to find your meal when a large boat with the loudest motor rides on the peak of the water. You and plenty of fish around you get thrown around from the humongous wake. Well, that

is what most sea turtles and other marine animals have to go through each day. I want to help solve this problem with the ¨Streamlined Shaped Boat. This boat helps by lowering the disturbance of other marine animals in the ocean. And I will

tell you how, but first let me tell you about my model for this product.

My product is based on the Green Sea Turtle's shell. The shell of a turtle is made of bone and cartilage like parts of your ear. The shell is covered with thin plates called scutes. Turtle's shell is made of 59 to 61 bones. These bones are made of keratin, like our fingernails. The turtle can not climb out of its shell because the shell is attached to their spine and rib cage. The shell's top is called the carapace, and the bottom is called the plastron. Turtles can feel pressure through their shell like we can through our fingernails. Sea turtles have a rounded, flattened, carapace and the entire shell is covered with tough, leathery skin supported by tiny bones. The shell's bone element are reduced, making the

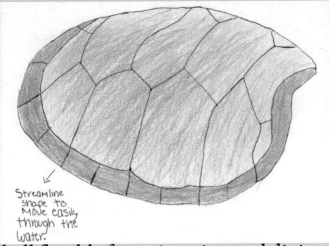

Streamline shape to move easily through the water.

shell flexible for swimming and diving.

You may be wondering how does my prototype work exactly? The streamline shape (based off the turtle's shell) helps with a small wake. How? Well most boats have a pointed bow and a horizontal back leaving a big wake. A streamline shape is one that is narrow at both ends, and broad or wide in the middle. A wake is the water that runs behind the boat that is made by when the boat moves. As the current reaches the boat front, the current spreads around the boat, so that leaves a large wake causing animals in distress. With the Streamlined Shaped boat the wake wraps around the boat leaving a small wake leaving animals in comfort.

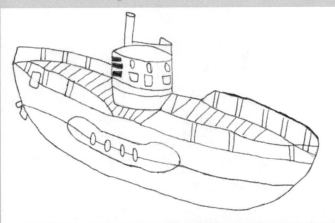

My prototype is very important to the problem because animals each day die because of boat propellers. The boat prevents extinction. Using an animal as inspiration is better than the original design because the boat mimics a real marine animal. The streamline shape helps to cut air resistance which makes the boat go faster while creating less wake. The streamline shape enables sea animals to swim faster.. My boat also reduces noise of the motor because noise also disturbs marine animals. Noise associated with shipping has the potential to cause disturbance to marine animals, including the marine mammals. The main source of noise is caused by the engine.

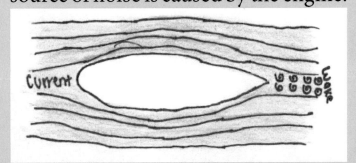

Current   Wake

In conclusion, I used the Green Sea Turtle as inspiration because they are very unique and are very strong. Also, my prototype is very important to humanity because the marine animals are getting killed each day by normal boats so my prototype will help with that and prevent extinction. If we have this boat the ocean will be a better place!

Regeneration Resolution

by Bryah I. Odom

Have you ever wondered if humans would ever be capable of regrowing lost limbs? With my product this is possible. For me, this topic has crossed my mind many times. A world where those who have lost their limbs because of health reasons, or fighting in the war won't have to live the rest of their life not being able to do the things they did before. That is one of the reasons why I decided to make the product I did. I want those who have suffered from limb loss and severe injury to be able to do they things they

> *There is an estimated 1.9 million people living with limb loss in the United States*

were once able to do.

The animal that I used for inspiration was the salamander.

There are many animals that are able to regrow lost limbs, but salamanders have the ability to regrow their lost limb almost exactly the same way that it looked before. For example, if

a Starfish' limb is cut off, it won't likely regrow looking exactly like the arm did before it was cut off. If you cut off the leg of a salamander it will grow back almost exactly the same as the leg looked before it was cut off.

My product is a medicine that will help people regrow or repair their body parts, or heal severe wounds. The process is quite simple. You take a shot that is made out of artificial fibroblast, a cell in connective tissue that produces collagen

> *My product is a medicine that will help people regrow or repair their body parts*

and other fibers. What the shot does is instruct the bacteria to extract the blastema from the

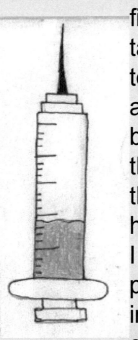 fibroblast and take the blastema to the wounded area. Once the blastema reaches the wounded area the it begins the healing process. I feel that my product is very important because there are many people that have lost their limbs because of health rea-

sons or because of fighting in the war. "Each day, more than 500 Americans   lose a limb. There is an estimated 1.9 million people living with limb loss in the United States," says the U.S. Department of Health & Human Services. That's a lot of people that are missing limbs. I made my product so I could give people hope that they don't have to spend the rest of their lives without limbs and being restrained from doing things they want to do.

The reason I chose the salamander is because salamanders have special regeneration skills that other animals don't have. My product is important because it gives people the chance to to do things they may have had to stop doing due to the fact they lost their limbs.

Saguaro Tower

by Ethan Perez

Saguaro cactuses are some of the most recognizable cactuses You may have seen them in an old western movie their distinct shape makes them easy to recognize. Droughts, there are three types of droughts I will be targeting two types of droughts hydrological (low levels of water in lakes and reservoirs) and meteorological (lack of rainfall) in Using the Saguaro cactus as inspiration will improve upon a drought situation by gathering and storing water

Flowers use photosynthesis to make food for the Plant.

Arms: Provide space for flowers to grow.

Stem/pleats: Stem has pleats which expand to store water until the next rain.

Roots: gathers water after a rain.

in it. So in case of a hydrological or meteorological drought there is water for the situation.

> "Photosynthesis usually happens in the leaves of the plant, but for the saguaro photosynthesis occurs in the top layer of the plant"

The stem of the Saguaro holds all the water for the plant, like all living organisms, a necessity for survival is food. Plants don't get their food from other organisms, instead they make their own food. This process is called photosynthesis. Photosynthesis usually happens in the leaves of a plant, but for the Saguaro photosynthesis occurs in the top layer of the plant. The Saguaro also has another adaption that we can't see, its roots. The roots gather the water after a rain and then it is stored in the stem. The Saguaro's roots gather water but where will the water be stored? Well, there is another part of the Saguaro called the pleats. The pleats are the creases in the Saguaro's stem. After a rain the roots do their job and gather water, but to accommodate this large amount of water the pleats expand like a pufferfish so it can store more water until the next rain.

When it rains on the land close to my prototype, the water will eventually soak into the ground, The artificial roots from my prototype, based off the roots from the Saguaro, will soak up the water. After the water is collected from the roots, it will be pumped it into the stem. The stem will be lined with sponge material so that the sponge can hold the water. This is modeled after the pleats on the stem. When the sponge material eventually ends up being saturated with water the hollow area can be used as storage too. This design is important because it can help regions in drought by supplying them with water they don't have and with water they need. We need something like this in the world because if a region is in a major hydrological or meteorological drought there will be water stored in my prototype in case of an emergency.

Using an organism as inspiration for my prototype is better than using an original design because my inspiration has made adaptations to survive in its harsh climate of the Sonoran Desert, and the adaptations that it's made can be an inspiration for a man made structure.

I chose this plant as an inspiration because it lives in the desert and can survive harsh conditions without rain because of the adaptations it has made. Again my design is important to humanity because it can supply water to people who don't have water.

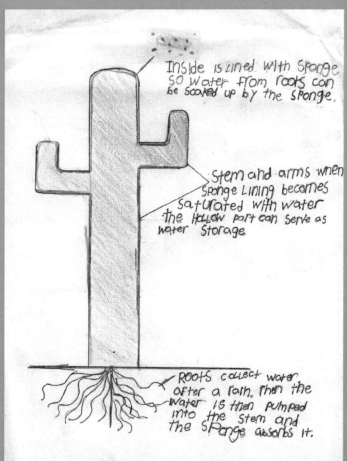

Inside is lined with sponge so water from roots can be soaked up by the sponge.

Stem and arms when sponge lining becomes saturated with water the hollow part can serve as water storage

Roots collect water after a rain, then the water is then pumped into the stem and the sponge absorbs it.

The Tiger Shoes

by Anaclaudia Uribe

What would you do if tiger boots were invented? Well, my prototype is called "The Tiger Hiking Boots". My boots would help hikers or anyone that just likes hiking in particular, because research shows that when you hike the improper lacing can cause your foot to slip downward and bang your toe into the front toe box and bruise. So, my prototype has a soft cushion to prevent that from happening. The tiger has a feature that helps specifically that, those are the soft and noiseless foot pads.

For my inspiration I chose a

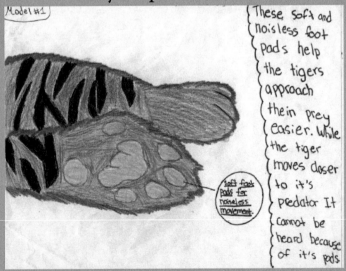

tiger because I'm really into big cats. My inspiration was when I went to Six Flags I saw this show about tigers

and I was really inspired because the tigers would climb up these tall wooden poles with rope tied around it. Their claws can dig into the pole so they can get a better grip. The tiger has retractable claws which means that at times their claws will be inside of the paw so they have extra space in their paws, so he/she can control how their paws go inside and how they come out.

My prototype works by helping all of the hikers that slip and fall. My prototype has retractable claws which dig into surfaces, It has a colorful tiger coat so people can see you and animals can't, and It has soft foot pads so neither people or animals can hear you. You can see the claws and soft pads on the bottom of the shoe, and you can see the tiger coat on the boots. The retractable claws have to go in and out of the boot so for my prototype the hiker would control it.

"I was really inspired because the tigers would climb up these tall wooden poles with rope tied around it."

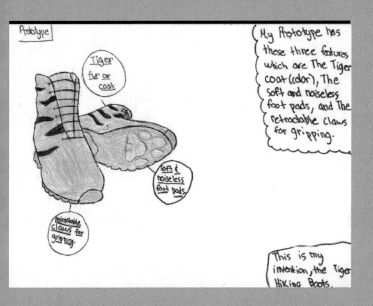

The hiker has a button on his or her hip that is connected to a wire which is connected to the retractable claws on the boots. They'll be able to press it and when they do, the retractable claws come out and the hiker does not slip or fall.

This prototype is important because it will cause less toe bruising and less slipping. We will also be needing this

for different reasons, for example we can also use them for the military. The military will be able to use these by getting the retractable claw to dig into the tree and they'll be able

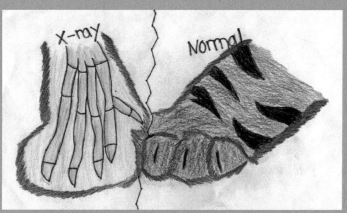

to climb up the tree so they're not spotted. Yes, I am aware that the Tiger Boots are bright and colorful. So, there would be a special type of tiger boots for the military, instead of the boots being orange and black they would be camouflage for them.

So in conclusion, I believe my prototype is very useful and it will help the world in some ways. My prototype would be made out of normal hiking boots with a coat of tiger print, my retractable claws would be made out of anything that could dig into the uneven surfaces like wood, metal, etc., and my soft foot pads would be made out of memory foam. Also the military would have the same features except instead of the colorful tiger print they'd have camouflage print.

Steer Fins

by JT Smith

Scientists have found out that the fastest Mako Shark in the world was caught on tape and it's speed reached a whopping 50 mph! My product improves the standard wetsuit by adding two pectoral fins to each side of the person.

Scientists have found out that the fastest Mako shark in the world reached 50mph

Sharks are the true king of the sea. Mako Sharks are the fastest shark of the bunch. Mako Sharks have a torpedo shaped body and a very long caudal fin (aka tail fin). Both the Mako and the Tuna have this body design. Mako sharks have a long and flexible spine which allows the shark to move it's caudal fin very fast which means it can swim fast. When I learned all of this information I thought, "Why can't we move so well in the water?" and researched how pectoral fins worked. Pectoral fins are located on the both sides of a shark. All the information I gathered from the web page, SharkSider, gave me the idea to make the steer fins!

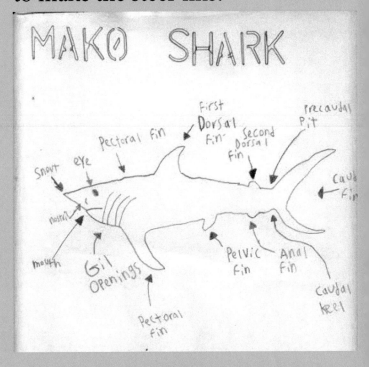

Steer fins are pectoral fin shaped fins that can attach to any swim suit or diving suit. To attach the fins, clip the two metal "attachers" on to each side of your body.

The "attachers" are made of aluminum magnesium alloy. Aluminum is a very malleable metal which means it's easy to bend it. So, I decided to infuse it with magnesium. Magnesium is a very strong material but at the same time very lightweight. Both materials combined create a very strong alloy. The fins are made up of cartilage like material and aluminum so it will be very flexible. Once you put on your fins and you enter water you will notice that turning is a lot easier. You will also notice how easy it is to swim because the fins allow you to glide through water like a knife through butter. Well, maybe not that easy. But, it should be a breeze.

I thought this was a very important product because when I put on a diving suit or wetsuit, I feel restricted and tight. And it doesn't improve anything except to keep you warm. I felt like I needed to create something that improves speed and agility. The pectoral fins are an important part of a shark, as it helps with turning, overall speed, and keeping the shark from sinking to the ocean floor.

In conclusion, the Mako Shark has many qualities we should take advantage of. Such as its unique muscle structure for people with weak muscles, muscle cancer, etc. Maybe it's torpedo body will be used for planes and trains. These fins will definitely help many divers and swimmers across the globe with swimming or even the Olympics! It could possibly help researchers and scientists. I personally think it will save a life.

STEER FINS

The Mantis Shrimp

by Elijah Douglas

The Mantis Shrimp have eyes that can move independently. "It uses this exceptional eyesight to avoid predators and track down prey." (National Aquarium). The Mantis Shrimp can see ten times more color than humans can that is including ultraviolet light. Mantis shrimps also have 16 color receptive cones to a human's 3 receptive cones which was believed to allow the Mantis Shrimp to see colors that are unimaginable to humans.

The goggles are made out of light, bulletproof titanium plating and the lenses are made out of a bulletproof glass. My prototype is influenced by the mantis shrimps movable eyes and color receptors, as well as the mantis shrimp's ability to see polarized lights and circular polarized lights. This will allow my prototype the ability to see objects and movements that the human cannot see. For example, you could rip a mantis shrimp's eye out and it will still maintain full depth reception in the other eye. My prototype is designed to operate despite damage to one eye.

When you are wearing the mantis shrimp goggles you will see the way a mantis shrimp sees the world. You would be able to see in ultraviolet, infrared, and you would also be able to see polarized lights, circular polarized lights, and colors no one else see or imagine.

To activate the independent movement control over the lenses there are three buttons on the side of the goggles, each one is a different color. Button 1 is to switch through vision modes and the color of button 1 is red. Button 2 is to activate and control the left eyes independent movement and the color of button 2 is blue. Button 3 is to activate and control the right eyes independent movement and the color of the button 3 is green.

This prototype is important because it helps the military with their stealth operations. The Mantis Shrimp's adaptations are important because it helps

this animal survive. With its independent eye movements, it can see predators from different angles. By using this adaptation in my goggles, military men and women will be able to see enemies or danger from different angles, just like the mantis shrimp. They will be less vulnerable to surprise attacks from behind or the side by looking through the goggles.

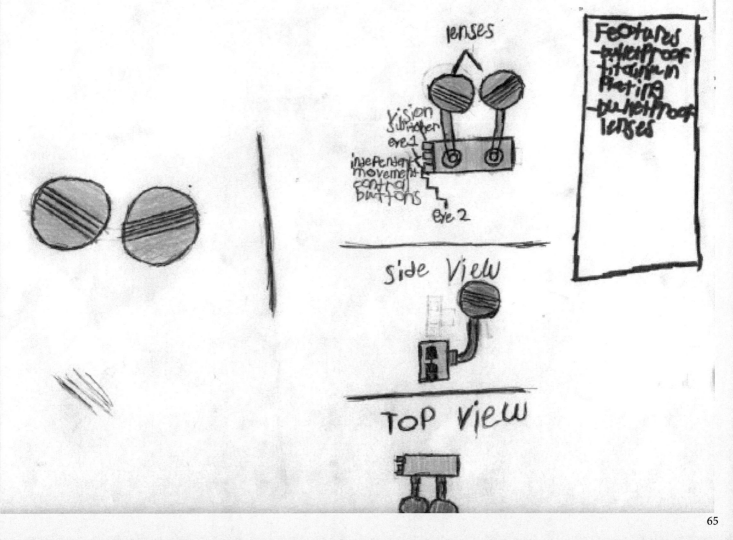

Puffer Tire
by Leeanye Wade

Did you know that in 2003, about 414 people died in accidents because of a flat tire? I believe that my prototype can fix this. My prototype can stop many people from getting flat tires and stopping some people from getting in car crashes lowering the amount of people dying from accidents.

and make it harder for it to be damaged. Also, it has a very stretchy stomach that can

"An estimated 414 fatalities, 10,275 non-fatal injuries, and 78,392 crashes occurred annually...my design can stop the cause of some accidents."

The puffer fish is able to suck in huge amounts of air and they have very stretchy stomachs.

My project is based off of the puffer fish. They have spines on their skin, also known as their spikes. The spines help protect the skin of the fish

hold huge amounts of air or water. This is how they puff up and

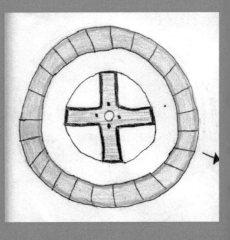

this makes them more durable so when they get attacked it will be harder for them to get hurt. The spines only stick out of the body when the fish has been threatened and it inflates itself. They usually can't be seen because they lay down on the body and are the same color as it.

The prototype I am making is an extra layer of material, put on the inside of the tire, made of pockets, that when touched they suck in air and they puff up blocking off the hole. The air is not able to come out and the tire is still filled up to where it is puffed up and able to be used until a new tire is bought. This layer of material is only on the inside so it is not shown and unable to get ruined. The spines within the tire will not show on the outside of the tire. This is important because, "According to a 2003 NHTSA report, an estimated 414 fatalities, 10,275 non-fatal injuries, and 78,392 crashes occurred annually due to flat tires or blowouts before tire pressure monitoring systems (TPMS) were installed in vehicles." http://www-nrd.nhtsa.dot. gov/Pubs/811617.pdf. So, I believe that my design can stop the cause of some accidents.

I chose this animal because it has special parts that others don't have. Its parts are more protective and stronger, so I think that it can help with building things. My prototype is important because if this works out it can help stop people from getting in accidents.

The Pitcher Plant

Lily C.D.

When do you notice when your customers stop coming? It's when your building doesn't look nice and professional. My solution will blow your mind! The pitcher plant it is slippery when exposed to water. Inspired by this adaptation of the plant, I designed a prototype that will make things slip right off a building, both solid and liquid, so it will look clean nice and professional for your convenience.

I chose to use a pitcher plant as inspiration. A pitcher plant is a plant that traps its own food without even moving! The the pitcher plant is named after a pitcher for lemonade because it looks like one. The rim is non-slippery when there is no rainfall or humidity, but when it is exposed to water the plant can be very slimey. Inside there is liquid nectar that is very tempting to drink to mice, lizards, frogs, and insects. When the insects slips inside the plant because of the slippery surface, it drowns in the liquid that it once tried to drink. The bodies dissolve into the goop and turn into nutrients for the plant.

My solution is very easy and inspired by the pitcher plant. First, you choose the spot that you need to protect, then you spray on porous teflon nanofibers less the one centimeter thick, add water, and oil-repellent lubricating liquid. After the wall dries add another coat. When the people do graffiti on the walls it will slip and when birds poop on your wall it will just come off. This is very important because vandalism is illegal and should be stopped. So after a few tries the taggers will get tired of wasting paint and will stop.

This invention will not only benefit you it will help others in their life.

I chose the pitcher plant because

it inspired me to help others. This will teach them a lesson to stop doing something that will keep staying the same. They can give up on their bad habits and start good new ones. While you have a great business. Their life will be patched up in some ways.

Revolution

by Alexander Servin

How do sharks swim so fast? This is a very interesting question about my project. In my project I designed a car that uses the aerodynamics of a shark. This prototype is going to change the car game in so many ways. Aerodynamics lets the car flow through the

> *"Aerdynamics let the car flow through the air more effciently"*

air more efficiently. The car would have a pattern all over the body. The pattern is similar to the placement of the scales on sharks. The idea is to have a similar body style on the car and use a pattern to have a more MPG (miles per gallon).

Sharks use patterns on their skin to move faster through the water. They have plenty of scales in the shape of teeth. The car is going to be using the pattern all over the body, which is going to make it more aerodynamic. That's going to help it flow through the air. Just like sharks, it reduces drag and helps them move faster. Their body shape with curves helps with getting their prey.

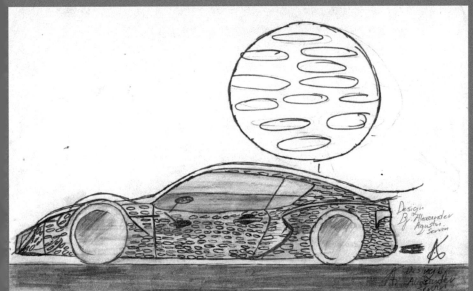

The details that go into my design are adaptations that the shark uses to move more quickly through the water. It has a very slick design which looks nice and very aerodynamic. The car will be very light because of its carbon fiber body finish. In my prototype you'll be able to see how much detail the car has put into its aerodynamics.

This design is very important because we don't have a lot of aerodynamic cars that look this great and that has so many things that it reflects on. It's great using a shark as inspiration because they're one of the most aerodynamic animals. They are so aerodynamic because of their scales. Sharks have been evolving for billions of years to become better and faster swimmers. Designing a car off of the shark's aerodynamics means that the car's design has billions of years of evolution behind it.

I used the shark as an inspiration for my car prototype because of it sleek edges and its scales. And I just thought, that a shark looked like a super car. My prototype is important because it would save money on gas, it would save oil and you'll look good driving such a car.

Chipmunk Cheek Medicine Bags

by Katherine Siefert

How many items do you think you could stuff in your cheeks? Well, the chipmunk can fit up to eight peanuts in the shells and one between it's teeth, or thirty-one peanuts without shells. Can you imagine carrying that much around in your mouth?

Chipmunks are expert hoarders. They collect items that do not decompose; like nuts, seeds and cones. They then take them to their burrows underground, which are complicated systems, running under logs and stones, or delving several feet underground. They can also be up to thirty feet long. Chipmunks have a main 'cache' (store) of winter supplies and then a few smaller ones in several hiding places across the burrow. If the chipmunk forgets where the hiding place is, he can use his sense of smell to find the hiding place. National Geographic states: "If the cache never is used it stays hidden until it germinates and becomes part of the woodland."

Chipmunks have a main 'cache' (store) of winter supplies, and then a few smaller ones in several hiding places.

Making food stores reminds me of the situation in Africa, and not having enough medicine stores. Hospitals and clinics in some African countries lack basic equipment and do not have enough medical supplies. The efficiency of the health systems depends on where you live in Africa. In the cities, you will probably get better treatment. But in the rural areas, they don't have access to basic needs like clean water or proper sanitation. This leads to diseases becoming fatal. Our Health-Africa states: "Half of Africa does not have access to the essential drugs they need to help people fight illness."

My idea is to use the chipmunk as inspiration to make a bag with pouches made out of strong, stretchy material to carry and store medicines. The stretchy cheeks of the chipmunk will be interpreted by the Bamboo fabric, which is strong, stretchy and, it's a safe **car**rying material for delicate objects. This bag will have multiple pockets for the medicine that is less delicate. The medicines that can't be handled as much will be put in vials that are inside the bag, which protects them from being harmed.

In conclusion, most places in Africa lack medicines that are needed, but my product can help by having a lot of the medicine in one bag. This product can be great for places in Africa that don't have as much access to the medicine they rightfully need.

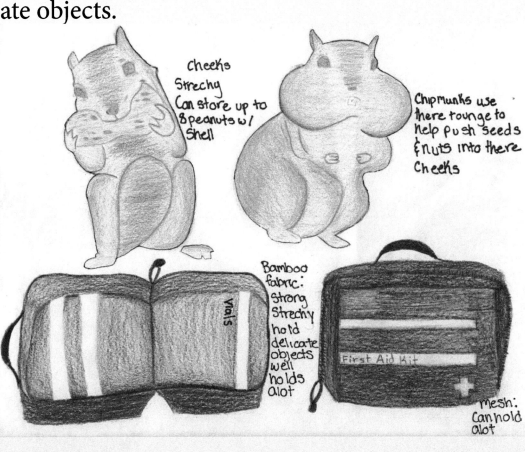

Cheeks Strechy Can store up to 8 peanuts w/ Shell

Chipmunks use there tounge to help push seeds & nuts into there Cheeks

Bamboo fabric: Strong Strechy hold delicate objects well holds alot

Vials

First Aid Kit

mesh: Can hold alot

Tech Glasses

by Miu Culham

Did you know that Leaf Nosed bats use sound waves and echoes called echolocation? They use echolocation to catch prey and they also use it when they are flying at night and can't see. *Imagine using the same technique as bats for people that can't see and improve their walking.*

My invention does just that.

The inspiration for my invention is the way bats hunt. When bats hunt or fly at night they use a technique called echolocation.

Bats emit sound by their mouth and nose as they fly. Bats make these noises by moving air past their vibrating vocal cords. They make these noises to hunt their prey and when they are flying at night.

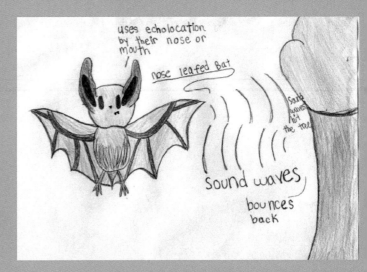

When the echoes bounce back to the bats ears, they can get a good enough picture in their brain so that they know what is in front of them. Another way they can hunt is by hearing. They can hear as far as 40 feet away.

My invention is tech glasses. The glasses will look like normal regular glasses/goggles. It will have a little sound box on the glasses that will send out sound waves. When the sound wave hits the object the box will automatically know that there's an object in front of them. It will then signal the brain to let the user know that there is an object or person in front of them.

These glasses could improve the world and help people walk again

The Tech Glasses could help for people that can not see. It could help people walk by themselves or for people that can not see well.

I chose a bat for this because I knew that bats use echolocation and I felt like maybe echolocation glasses could be a product or an invention that would be invented in the future. I know that it can change the way people that have trouble with their vision can get around.

Sends out sound waves sound box

Sends sound way when it hits an ob Sends signal to th

Mountain Goat Climbing Shoe

By Jasmine Ruiz

Have you ever thought of how much danger that you would be in during mountain climbing and hiking? Well, today I am going to tell you about the prototype that I designed. I was inspired by an animal that you may know, it goes by the name of mountain goat. The reason why I was inspired by this animal is because when they climb they have great balance. Due to the pads that are located at the bottom of their hooves, they have great grip so that they don't slip or fall. I think that this could be a very good solution for mountain climbers and hikers.

I think that it is important that we solve this problem because according to an article called "Statistics for Climbing Accidents, Injuries, and Fatalities" that I have read it stated that from "1998 to 2011 the RMRG [Rocky Mountain Rescue Group] rescued 2,198 mountain climbers and wilderness victims in boulder country. Rock climbing were 428 or 19.5% of all victims." I think that 2,198 is a lot of people that needed help from a rescue team. With this prototype it will not only help hikers and mountain climbers but it will decrease the chances of danger that they are in.

> "2,198 is a lot of people that needed help from a rescue team"

When I did my research I learned that some deaths occurred when they were climbing and didn't have great grip so they slipped. Due to the body weight of the climber, the person holding the rope at the top released the rope and the climber plunged to their death. Just to make it even more safe I will make a slot for each toe to individually go in. This was inspired by a mountain goat because the mountain goat has two toes that can spread wide, this results in great balance. They also have pads that are located at the bottom of each toe and that help them have grip. I will add no laces so that just in case when they are climbing they won't have to go back down just to tie their shoe. I will also make the shoes lightweight so that when they are climbing they won't have to carry extra weight.

My prototype is important is because it lowers the chances of danger that mountain climbers and hikers have. It will decrease some problems that they may run into. Imagine if a fellow cousin or broth-er/sister tells you I am going hiking with some friends and while they are in the process of hiking you get a call telling you that they slipped and fell and the result was a broken leg. What would you do? This is the kind of thing my prototype prevents. The mountain goat inspired me to think of this problem because when I read the article "Mountain Goat" it stated, "the mountain goat can jump 12 feet in a single bound without falling", I thought that it would be cool if humans could have the same ability and that is how the idea of my prototype came along.

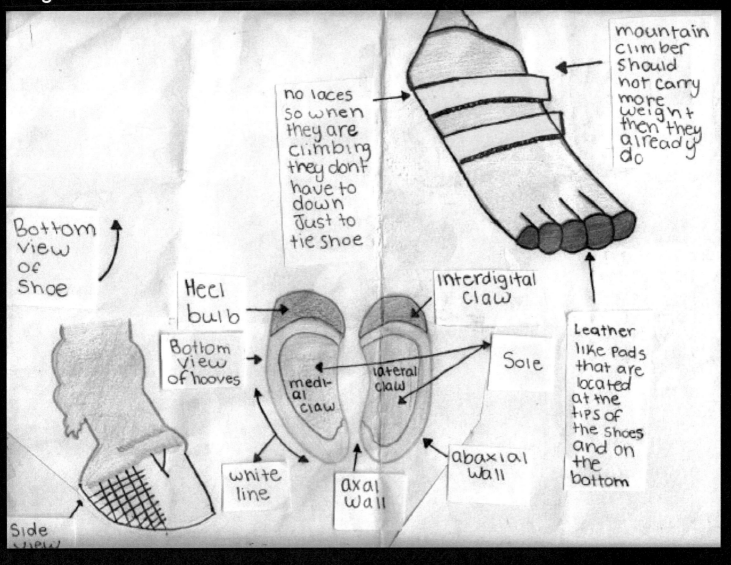

Glowing Eels

by Lucía Brown

Have you ever seen a glowing eel? Me neither, but they do exist in Japan. Quite recently it was discovered that the Japanese freshwater eel (anguilla japonica) has a new fluorescent protein in their muscle tissue that allows them to use the blue light from the moon to be absorbed and give off a neon green glow. This discovery could save lives by letting doctors easily be able to see if there is something wrong with your liver and to make sure it is working properly.

> *Too much Billirubin in your liver can lead to having Jaundies*

According to Biotechniques. com, there is a protein in the muscles of the anguilla japonica that glows when it connects to bilirubin. Bilirubin is released in the body when red blood cells are being broken down. This discovery has lead scientists to believe that this could open the doorway to many more discoveries of new

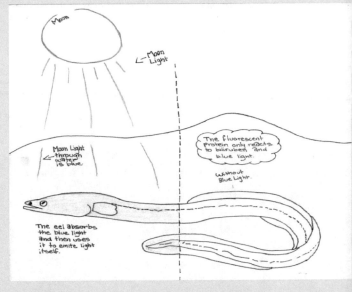

fluorescent proteins that could help scientists better understand and examine the cells that dissolve muscle cells.

For this product to work, the doctors would need to add the

protein to the liver they are trying to diagnose and the liver should start to have a faint glow (it is normal to have bilirubin in your liver), the problem is when the liver gives off a strong glow, that would indicate that there is too much bilirubin.

A new Fluorescent Protein has been found in a Japanese Eel.

The Japanese water eel uses the blue light from the moon to put into function the fluorescent protein. The eel uses the glowing quality to find mates in the dark depths of the river.

I think this product is important and can save lives. I think it would help doctors be able to tell if someone's liver is not functioning properly in a quicker way than having to scan the liver and wait for results. If there are problems the doctors can go right away to try to fix the problem.

This discovery could save lives by letting doctors be able to see if there is somethin wrong with your liver.

I chose the Japanese eel because I thought it was a cool animal and as it turns out it had something in its body that can save lives. Who would pass up the opportunity to spread the word about something that can save someone's life? Keep those livers happy and healthy.

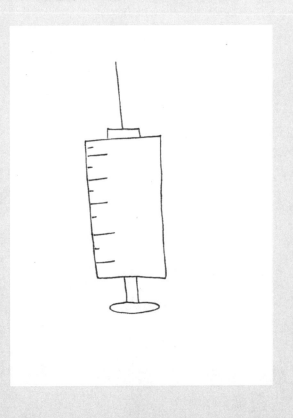

Light The Way

by Keren Quintana

In" The Angeles Times", an article written by Tomas Alex Tizon, he states,"Richard Hills was one of 3,323 people reported missing in the state last year, not a record but far higher, relative to population, than anywhere else in the country. On average, 5 of every 1,000 people go missing every year, roughly double the national rate. Since Alaska began tracking the numbers in 1988, police have received at least 60,700 reports of missing people." Crazy, right? I'm trying to change that with my invention called Light the Way. Light the Way is a robotic firefly,except that the firefly will be twice as big. This robot will be able to find lost people faster than the police.

> # 5 of every 1,000 people go missing every year

Each species of firefly has its own pattern of light flashing. Fireflies emit light mostly to attract mates, although they also communicate for other reasons as well, such as to defend territory and warn predators away. In some firefly species, only one sex lights up. In most, however, both sexes glow; often the male will fly, while females will wait in trees, shrubs and grasses to spot an attractive male. If she finds one, she'll signal it with a flash of her own. Some fireflies are bioluminescent, meaning that they can produce their own light. When attacked, fireflies shed drops of blood in a process known as "reflex bleeding." The blood contains chemicals that taste bitter and can be poisonous to some animals. Because of this, many animals learn to avoid eating fireflies. Pet owners should never feed fireflies to lizards, snakes and other reptilian pets.

Light the Way, will look like a real firefly, just bigger and with two different color lights, front and

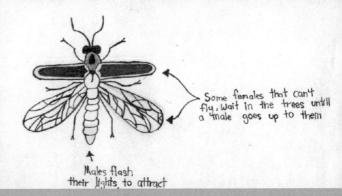

Some females that can't fly, wait in the trees untill a male goes up to them

Males flash their lights to attract

back. One color will be yellow, that

will be in the front, it will be for searching people in the dark. The other color will be green, that will be in the back, it will be to communicate to the other robotic fireflies. The robot will be able to fly really fast and find people by their heat, like snakes do with their heat sensors. First, it will take a picture of the missing person and keep it in the memory chip until it finds the person. It will scan the person from head to toe to see if that person is a match.

Light the way is really important to the world because we really don't want people to get lost

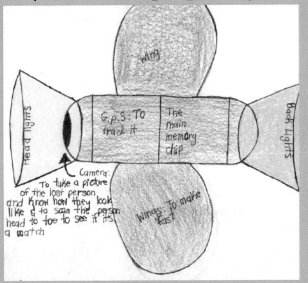

or go missing. It also takes the police a long time to find them. Now, it's going to be fast and simple for the police. The robotic firefly just needs a picture of the person and their work is done. Night goggles and clues don't always help the police to find a lost person. A firefly can be spotted really quickly , even from far away. We can make

a robotic firefly to find people by putting a gps, a camera, wings and a memory chip in it. This can save a lot people's life. On the website PInow.com it said that, ¨More than 800,000 people are reported as missing and are entered into FBI's National Crime Information Center (NCIC) annually. Of these, 85%-90% are minors. These statistics do not include those who are unofficially missing such as those who have not been reported as missing persons.¨

A firefly can fly really fast and it can glow from it's bottom. Male fireflies flash to attract females, but the flashing in the robot will be to communicate. We just need to make them a bit bigger to make it fly really, really fast and make them communicate to the other robotic fireflies. This can change the world into a better place to live because we will feel safe knowing ¨Light the Way¨ can find you. Just imagine a world without bad people that want to kidnap you. You will be able to walk in the streets or in the woods without worrying about people not finding you.

The Monkey Shoes

by Isaiah Thomas

Wouldn't it be awesome if you could climb on trees like monkeys do? That would be amazing wouldn't it?? Well, that's what my prototype will help you do. By researching monkey's feet I've seen how much they look like human hands and that's what I'm going to show you now makes

Orangutans are some of the best climbers in thier species

the design so efficient. The design is shoes modeled after monkey's feet.

I have been looking at the the Orangutan because they are some of the best climbers.

So, I am using them for the design of their feet. Orangutans have unique adaptations to their life in the treetops: feet designed much like hands for climbing, flexible hips for holding on in any direction, long arms for reaching and long, strong hands and feet. So they can climb very fast and the prototype enables humans to be able to do that.

The prototype is on the bottom of the shoe. It turns on based off of the motion in your foot. So as if you were making a fist in your hand, do it with your foot. As soon as you do that, the device on the bottom of the shoe will grab onto anything under it. And also adapt to it if you're stand-

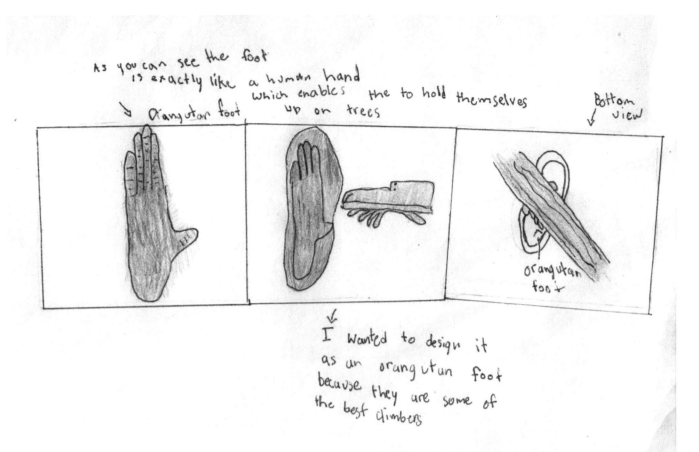

As you can see the foot is exactly like a human hand which enables the to hold themselves up on trees

↳ Orangutan foot

Bottom view ↓

orangutan foot

↓
I wanted to design it as an orangutan foot because they are some of the best climbers

ing straight it will grab from the front, and if you're facing the side it'll grab it from the side.

I think we need this to help people discover new thing. The apes can do some amazing things and were formed in a way that is more efficient than humans, in climbing ability. That's why they can climb so well and do multiple things at once because they have hands for feet. I think that is a key idea in the prototype, because then they can do more things than usual.

As I said before monkey's feet are shaped like hands which enables them to be able to climb and hang This would be a good idea for humans with the shoes that I want to make for my design because it allows humans to do things they are unable to do with their regular feet or shoes.

Mako Boat

by: Reagan Bosshardt

Imagine you are the fastest shark in the ocean gliding through the water. Well, that is how the mako shark feels. I would like to take these amazing adaptations and design a boat used by the Navy. It will help the Navy because they spend so much money on gas.

I chose the mako shark because the animal has very helpful adaptations such as its scales and it does not let algee stay on it. The skin of the mako shark has little

ridges so bacteria, plankton, and barnacles don't get stuck to it. One of the other adaptions is there wicked speed! Did you know that the mako shark can swim as fast as 40 mph? These adaptations gave me the idea to design a boat.

The boat I designed has mako shark-like skin with the speed of 120mph (3 times the speed of a mako shark!). The speed will help the Navy get to things faster. I also put the shark-like skin on the bottom of the boat so algae and barnacles will not stick to it.

The sharks skin has little ridges so barnicals don't stick to it

This boat is important because it will save millions of dollars a year. Did you know that the Navy spends millions and millions of dollars every year on gas? By having this boat it will lower the prices, because when there are barnacles on the ship it makes it much heavier. With the ship being weighed down it needs more gas. Those problems would be fixed if we swapped the Navy boats for my boat because my boat has the shark skin technology so the barnacles will slide right off!

In conclusion, I believe that with all of the money that is used for the Navy could be used for other things like schools or after school programs. With all of the amazing adaptations from the mako shark this boat will change the world!

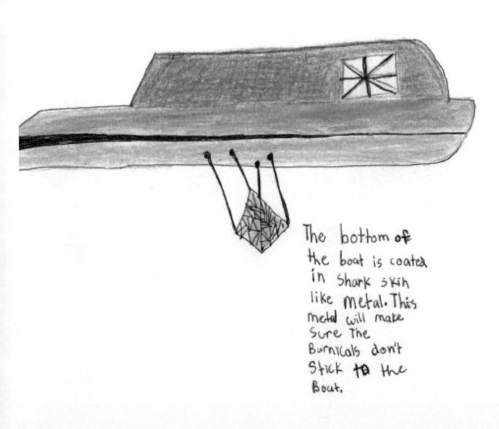

The bottom of the boat is coated in Shark skin like metal. This metal will make sure the Barnicals don't stick to the Boat.

Glowing Roads

by Kayla Manguil

Have you ever gotten into a car accident in the rain? Well, I have a new product that could help solve that problem. I have created a water activated road paint, so that when it rains, the road will glow so drivers can see where they're going.

This product was inspired by glowing algae. The light shows depending on the way the water moves around it. In the rain, while the water is swooshing around, the light will start to shine. Algae glows a blue-green color when the water around it is moving. Because the rain will keep dropping, the water will keep moving, thus the algae will continue to glow. That got me to thinking about how I could come up with something that could help the world somehow with using the glow from algae.

That's when I thought of water activated glowing road

paint! The paint will glow when the water from rain moves around, that is most likely considering when in rains, puddles are formed as more rain drops in. The algae can grow anywhere where there is water, carbon dioxide, sunlight, and minerals. The road would be a great environment for the algae to grow and glow.

> *According to the Federal Highway Weather Administration, 46% of car crashes caused by weather are because of rain.*

This product is can be very useful when it comes to driving in the rain. According to the Federal Highway Weather Administration, 46% of car crashes caused by weather are because of rain. That's a lot! This product could help people drive in the rain and help bring that percentage down. It could help by letting them know where the road leads to and where they're going with the glow. It could also be useful at night in the rain because you most likely don't know where to go or if anyone is in front of you or behind you. This new road paint is better than regular road paint because it could help you be safer while driving.

I used glowing algae as my inspiration because it's fascinating to see how it glows. As you can see, this product could be very useful when it comes to driving in the rain and/or in the dark.

Hummingbot

By: Matthew Mansfield

"From 1992 to 2007 there were 78,488 individuals involved in 65,439 Search and Rescue (SAR) incidents."

"From 1992 to 2007 there were 78,488 individuals involved in 65,439 Search and Rescue (SAR) incidents. These incidents ended with 2,659 fatalities, 24,288 ill or injured individuals, and 13,212 saves. On average there were 11.2 SAR incidents each day," says Jerrie Dean, founder of Missing Persons of America. Underprepared hikers wind up lost all of the time. It can take very large amounts of time and people to search for a lost person, and the longer it takes, the more the lost victim will get hungry, thirsty, hot, or cold. My idea, the Hum-mingbot, could help find hikers in an easier, and shorter fashion. The Hummingbot is a hummingbird-like robot with the ability to help find lost hikers with quick speeds and heat sensing abilities.

I chose the hummingbird as the inspiration for my design because it has some of the best maneuverability of any bird. The hummingbird can fly right, left, down, up, and surprisingly even upside down! They can also hover in place. Average hummingbird speeds are around 25 to 30 miles per hour. All of these abilities could be used to find a lost hiker faster without having to send out large search teams.

The Hummingbot is an automatic system, except from a few commands the

Live videos are shown here

Four bots will be sent out and each provides a live video stream of what the cameras are seeing.

Each bot spends around 30 seconds on each heat signature. Any bot can be commanded to watch a heat signature for a bit longer.

Commands

- Move on= Move to a different heat signature.
- Return= Return to launch area.
- Stay= Stay in current area for 10 more seconds.
- Drop= Drop GPS beacon.
- Speak= Activate speaker.
- Flashlight= turn flashlight on or off.

operator can activate. It is equipped with a camera, speaker, heat detector, locater beacon, and a flashlight. Four or five bots would be sent out to find the lost hiker. The camera would provide a live video stream of what the Hummingbot is looking at. The heat detector is used to attract the Hummingbot to a heat signature around 97.8 degrees fahrenheit (average human body temperature), so that it can actually find the victim, rather than just aimlessly wander around looking for the victim. The locater beacon is dropped once the operator overlooking the live video streams from the Hummingbots has confirmed that the heat signature the bot found was a human and not an animal.

Hummingbird's flight pattern by beating it's wings very fast in short strokes to allow the right amount of lift for its small figure. Hummingbirds flap their wings by twisting the humerus. They have the ability to produce lift on both the downstroke and the upstroke by rotating its wings backwards due to its flexible shoulder joints that allows a rotation of 180 degrees. The Hummingbot will mimic all of these attributes of the Hummingbird to allow the best possible mobility.

The Hummingbird was definitely the right choice for my idea with its quick and nimble movement. The Hummingbot is a faster alternative to search teams and

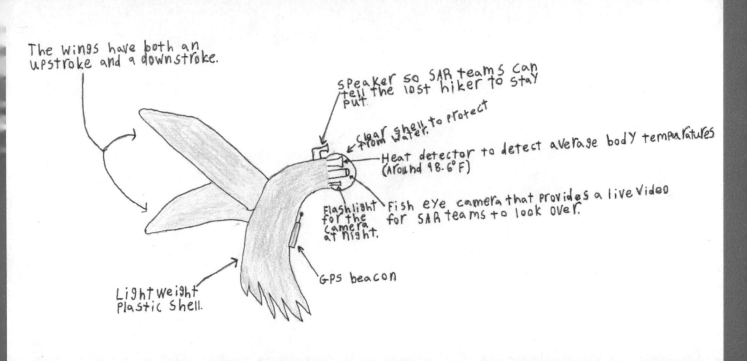

It is dropped by the operator by activating a command that drops the locator beacon so that rescue teams can track the victim.

The Hummingbot will mimic the

drones, because it can move at fast speeds while being very agile. If the Hummingbot is used by rescue teams it could help lower the number of hiker injuries and casualties.

Mako Shark

by Anthony Asciutto Matkowski

What is interesting about the Mako shark is that its skin it not flat like it seems it is choppy and rough. The Mako shark can reach up to speeds of 35 to 45 miles an hour faster than most motorboats.

The animal that inspired me was a Mako shark because it was

"The shark skin is super choppy and rough"

sleek and fast and could beat most motor-propelled boats. Its adaptations have given it this type of skin that is not flat but choppy and rough so it creates whirlpools that push it through the water. Evleshin are what makes the shark fast.

The way my Sharkskinz wetsuit is made is by copying over the texture and skin of a Mako shark and put it onto a wetsuit. The way you can see the adaptations of a Mako shark is in the scaly skin of the wetsuit. It uses the same type of science from the Mako shark to create whirlpools on the wetsuit to push the swimmer along through the water. The reason why we need the sharkskin wetsuit is to help travel fast underwater with a less breath.

This can help the Navy and could potentially be used for boats by using the same type of Sharkskinz for the hull of the boat.

The reason that I picked the Mako shark is because it's such an amazing animal. It has so many things about it that we don't even think of when we see it. The reason that my prototype is important is because it will help the navy to swim fast under water.

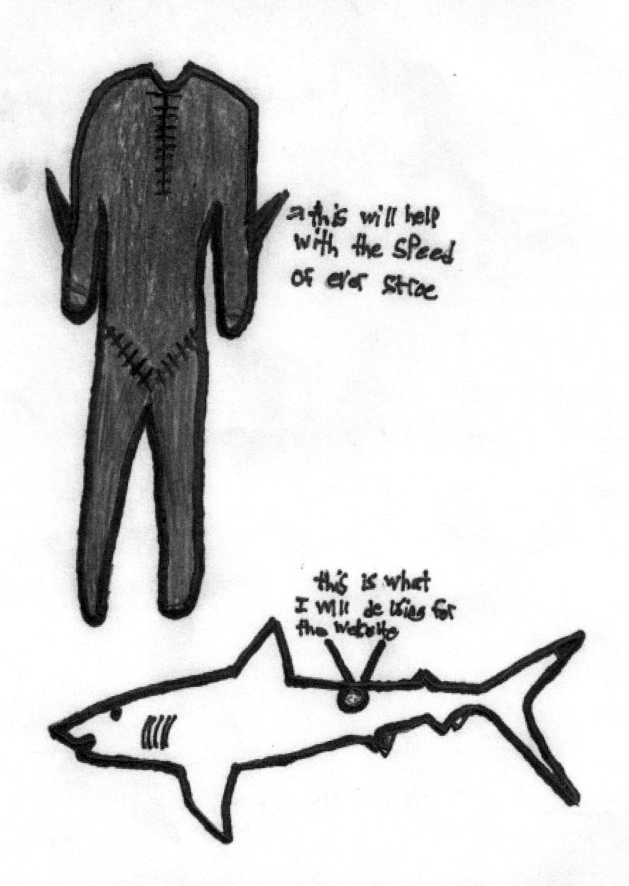

Snow Hare Suit

by Alexander Sammuli

The snow hare can change its fur color with the seasons. My design is a suit that is inspired by some of the adaptations of the snow hare. It can help people in the Military or the NAVY by blending in with its surroundings, the suit changes colors with the seasons"." You might ask how it can change colors with the seasons, well there's a micro chip in the suit so when the seasons change the suit changes with it.

The snowhare can change different directions to escape its prey. Its fur changes with the seasons for camouflage. They are also great swimmers, the snow hare can live up to 3 years in the wild. Many predators including owls, hawks, lynx, coyotes, wolves, foxes, marten, mountain lions, bobcats, weasels, and humans. In the summer they eat greens, grass, and flowers. In the winter, buds, twigs and bark.

My suit helps people in the NAVY and the military so it camouflage with it's surroundings so the enemy can't see them coming It can also keep you really warm in the winter because it as three layers of coting. You can see it in the future in the

The layers in the suit.
(outer layer) Silk
Cotten
Warm Pad

It has a thick layer of Pating.

It has a Micro chip in the back of the Suit and When the Seasions change the Suit also changes.

they can camoflage with there Serou dings.

Marine Corps, NAVY, Military, etc. The original idea was a ghillie suit a ghillie suit is a suit that is a camouflage suit that is made of a green and a bright

brown so it blends in with bushes. Then I upgraded it to change in the sessions so no one can see the people that are wearing the suit coming

This suit is important because too many people get killed in war"". so that's why I'm making this suit"". It Can help people and some of my family members were in the marines, so that's why I'm making this suit. It's inspiring to use this animal and make something out of it because I want to save hundreds of people.

I chose this animal because it can save lives in war and it can help people when they are in war. I designed this because I won't to save lives and to help people in the war, military, NAVY, etc. It can make you warm fast so you don't get hypothermia and frostbite. You can stay in the arctic for a long period of time so you can survive out there.

(Fun Fact)
The snow hare can run up to 27 mph (43 kph). The snow hare can jump up to 3 more 10 feet in one jump.

Robo Legs

by Raul Velazquez

How do kangaroos jump high and move fast? I designed a mechanism that a person can wear on their feet and legs that is used to jump higher and move faster, just like a kangaroo. The need for my prototype is important because the kangaroo speed and hoping can help us not be late to work. It can be useful to cops for chasing suspects. It can be useful to the military for running or evading obstacles in the battlefield. It can save your life or it can save your job/career.

The reason the invention is needed is because people get late to work and this invention can prevent that. It can prevent people escaping police chases. If there was a cop chasing a criminal, but there was a lot of traffic, he can just run/hop to the scene of the crime faster. It can contribute in military operations. The prototype works by putting it on your feet and legs, then the latches auto latch on. After that, there's a small thin piece on the thigh part of the leg device that sends a message to the brain to move faster. The device burns a lot of calories while the user is wearing it, so drinking a energy drink can give you more energy to use it. It has springs on the foot part so it could bounce up in the air when weight pressed on. The device is made out of microlattice or it could made of alloy. Microlattice is light as styrofoam and strong as titanium.

We need this because people escape a police chase and the criminals that escape can be a threat to people. The leg device can outrun traffic and probably catch up to the criminal. We also need this because it can make the percentage of people late to work go lower. Another reason we need this is because the military can find it useful, it can evade or outrun a bomb or a car coming at a high speed. This design is probably better than prosthetics, but the prototype and prosthetics have their differences. Prosthetics just give you a metal leg if you don't have a leg. But the leg device has more speed, hops, and quick evading movements that can occur. The leg device can be used on people with legs and no legs.

The large and strong tail is used for balance when hopping, and as fifth limb when moving about on all four legs or five. A kangaroo uses their strong tail to balance while jumping so they don't trip. Kangaroos use their powerful hind legs to do great speed, their speed can go up to 35 miles an hour. Their bounding gait can allows them to cover 25 feet in a single leap and jump 6 feet high. They are the tallest marsupials, standing over 6 feet. That is why I chose the kangaroo, because it has great speed and hops. This leg device can help the world by, dodging a tragedy that would've occurred without the leg device.

My prototype helps people because the hop/speed can make you evade faster. Which means to escape or to dodge. My prototype can also help people by getting to something in time or going somewhere in time. Another way it can be useful is avoiding traffic. In conclusion, my design will help many people.

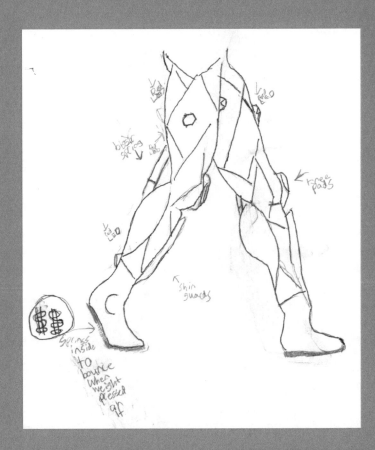

"This leg device can help the world by, dodging a tragedy that would've occurred without the leg device."

Modified Germ

by Andy Portilla

Did you know that the Hydnellum peckii (Bleeding Tooth Fungus) exchanges minerals and amino acids for fixed carbon? ¨Hydnellum peckii is a mycorrhizal fungus, and as such establishes a mutualistic relationship with the roots of certain trees (referred to as "hosts"), in which the fungus exchanges minerals and amino acids extracted from the soil for fixed carbon from the host¨. states Wikipedia.

The Bleeding Tooth Fungus exchanges minerals and amino acids from the soil to its host for fixed carbon. My prototype is a modified germ to attract carbon molecules. We could then extract that germ and have natural carbon. I think my prototype is useful because we could naturally make fuels, carbon fiber, graphite.

What scientists learned is that there are several greenhouse gases responsible for Global Warming. Most come from the combustion of fossil fuels in cars, factories and electricity production. The gas responsible for the most warming is carbon dioxide, also called CO_2.

I chose this organism because I want to stop global warming and make cars better and safer." states NRDC. Using an organism as inspiration is better in my case it would help stop global warming.

I used this organism because I saw a video on naturally grown batteries and I was so amazed by that idea I decided to use it for natural carbon. My prototype is important to humanity because it will help make the earth healthier by decreasing the number of factories.

"On the other hand, the bloodlike substance has anticoagulant and antibacterial properties. It's nature's next penicillin! All you have to do is lick it. Go ahead."

Glossary

Mutualistic: The way two organisms of different species exist in a relationship in which each individual benefits from the activity of the other. Similar interactions within a species are known as co-operation.

Amino Acids is a simple organic compound containing both a carboxyl (—COOH) and an amino

NRDC stands for Natural Resources Defense Council

Co2 means Carbon Dioxide

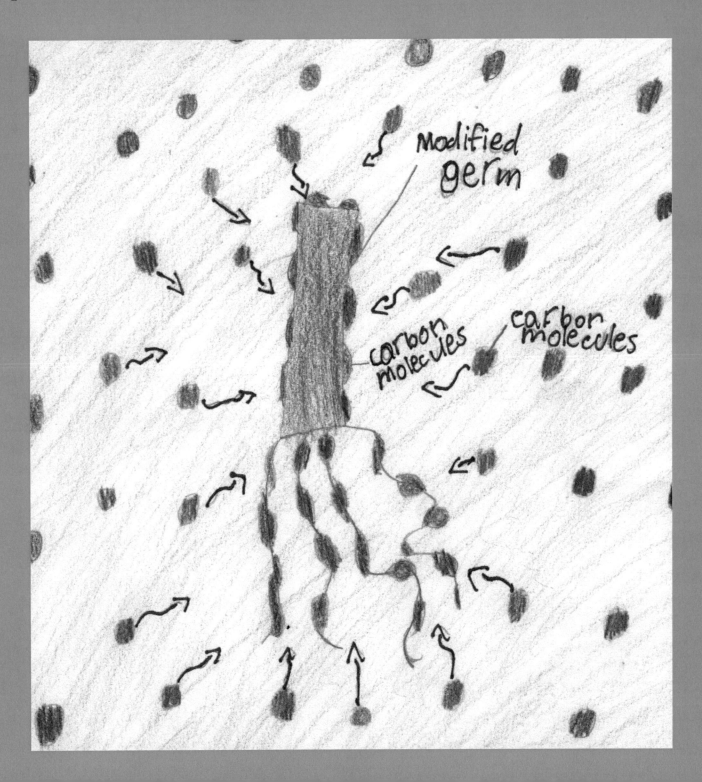

Healing Power

by Estrella Andrade

Did you know that axolot can heal themselves to the next level? About 1.8 million Americans are living with amputations. Axolotls can receate their whole body or part of their body that has been amputated. But I wonder if we had axolotl DNA in our body,

---

About 1.8 million Americans are living with amputations.

---

could we heal ourselves? In the future, we will turn the axolotl's DNA or stem cells into a pill so people can swallow the pill. With the axolotl you can cut the spinal cord, crush it, remove a segment, and it will regenerate. You can cut the limbs at any level—the wrist, the elbow, the upper arm—and it will regenerate, and it's perfect. There is nothing missing, there's no scarring on the skin at the site of amputation, every tissue is replaced. They can regenerate the same limb 50, 60, 100 times. And every time: perfect.

How the pill will works, is that it will heal the person that swallows the pill by healing their arm or leg that has been cut off

How the pill will works, is that it will heal the person that swallows the pill by healing their arm or leg that has been cut off and heal their broken bone. By using these pills, people that have lost a limb or were born without a limb, can have an arm or leg and a limb.

This design is important because, well if you put it like this what will happen if you had lost a limb? This design will help kids or adults that have lost limb or born without it.   We need it because if it were a lower limbs they can walk again on their own or use of their upper limbs. I chose this animal because I see men and women that have lost a limb and came back from war. These people showed us that they fought for our country.

It's important because they can have the ability to walk again. Also to use their upper limbs.

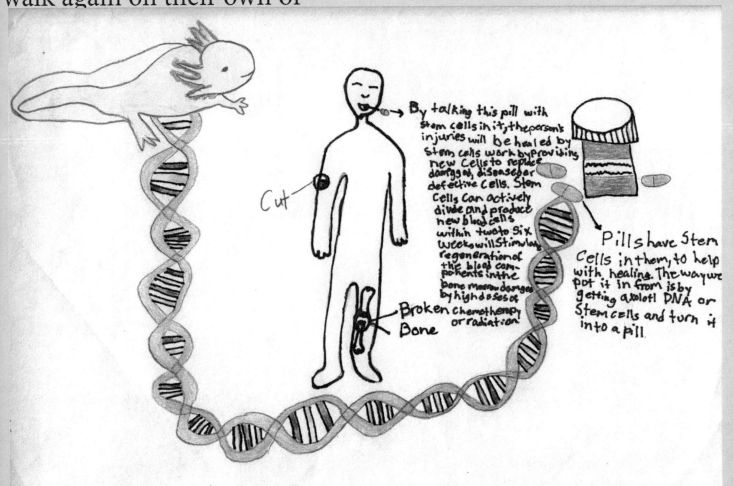

Cut

By talking this pill with stem cells in it, the person's injuries will be healed by stem cells work by providing new cells to replace damaged, diseased or defective cells. Stem cells can actively divide and produce new blood cells within two to six weeks will stimulate regeneration of the blood components in the bone marrow damaged by high doses of chemotherapy or radiation

Broken Bone

Pills have stem cells in them, to help with healing. The way we put it in from is by getting axolotl DNA or stem cells and turn it into a pill.

The Helm-ute

by Miguel Amador

The alligator. Take one glance at these frightening masters of massacre, they are armored, powerful, and it is obvious that they are descendants from the distant past. Scientists say that the species is more than 150 million years old, they managed to avoid extinction 65 million years ago when the dinosaurs died off. Alligators have armor called scutes or osteoderms. I'm going to design my prototype based off of the alligator's body armor and use it for the outer layer of a football helmet. It's important because there are a lot of concussions in football. We should change that.

Although abundant and ungraceful outside of the water, the armored giants are supreme adapted swimmers. Males average from 10 to 15 feet in length and can weigh up to half a ton, and the females grow to a maximum of about 9.8 feet. Both males and females have an armored body with a muscular flat tail. They have embedded bony plates called scutes or osteoderms. The scutes are bony and rigid plates that's why the skin is tough protecting it from enemies and predators. These are what I'm going to copy for the helmet.

The prototype is going to

"High school football accounts for 47 percent of all reported sports concussions, with 33 percent of concussions occurring during practice"- Sport Concussions

work because there are two layers of armor. The outer layer is going to be the synthetic scutes, then there's going to be a soft cushion under the synthetic scutes. There is going to be another layer of scutes for even more protection, finally the last layer is going to be a grippy soft cushion layer. On the Smithsonian National Zoological Park site it says, "the skin on the back is armored with embedded bony plates called osteoderms or scutes." That is what I'm basing the layers off of.

The design is important because the helmet has two layers of protection and comfort. So it's very protective, comfortable and the helmet won't come off. The world needs this cause it will reduce the risk of concussions and head injuries, and it will be a more comfortable helmet. I think the inspiration from the alligator is better than just the regular idea because the regular helmets only have a hard layer and there's not a lot of cushion in the helmet. In Sport Concussion Statistics it says,"High school football accounts for 47 percent of all reported sports concussions, with 33 percent of concussions occurring during practice" I'm choosing this animal because it is one of the most armored and dangerous predators in the world, and I felt this urge to study this animal. My idea is important because it will help with head trauma and it will be a more comfortable helmet. It is an idea to change everything mentally, and physically.

No Stingy Hurty Death

by Keegan Bollinger

Dangerous jellyfish that can kill you in less than two minutes? SOUNDS FUN! For all the people who think very dangerous jellyfish are cute, and possibly have a death wish, you may want to use this. Introducing anti-hemozoin spray. You simply spray it on your body/self before you go to hug the jellyfish! Boom! No stingy death! That's good.

Dumb wetsuits, not protecting you… They should really make jellyfish proof wetsuits.

So, am I done with this? Will I stop asking questions? No, I won't. Anyways, the Galactus Atlantis is a very interesting and rare sea slug. Now, you may think a sea slug is a "bottom feeder" that lives at the bottom of the sea. For the Glaucus At-

lanticus it's the opposite. It floats *It feeds on the very dangerous, Portuguese Man of War.* near the surface and moves with the currents. But what does this have to do with and anti-hemozoin spray? Well, I'm glad I asked

myself that. It feeds on the very dangerous, Portuguese Man of War. It can achieve this feat because it has anti-venom in its mouth.

This spray works as any spray sunscreen works, with a spray can. It takes the anti-hemozoin saliva and simply puts it in a spray can. Spray cans work by using matter that can quickly change from liquid to gas. For example, water and paint. Spray cans dispense a liquid under pressure. The pressure comes from a gas that is inside the can. This gas is called the propellant. In my anti-hemozoin spray, Galactus Atlantis saliva is the propellent.

This spray will help scientists capture, tag , and examine jellyfish. It can also help people get through jellyfish to help stranded whales and deep sea dive without getting hurt. The saliva from the galactus atlantis is able to protect it from the most poisonous jellyfish in australia.

In conclusion this spray will help scientists and civilians stay safe from jellyfish in a small and easy way.

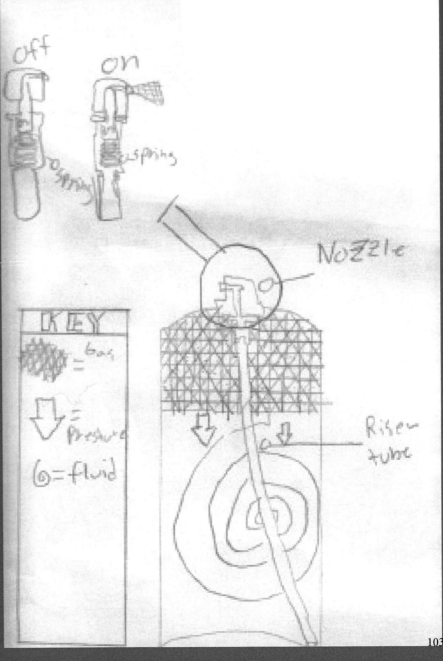

Blue Whale Whistle

by Benjamin Tamayo

The blue whales are the loudest animals when they call to each other. Their call can reach a little under 1000 miles. A little over 100 people get stranded at sea and have a slight possibility of surviving because of the extremely cold waters and the dangerous sea animals. But now when you board a plane you can get a whistle. It's a whistle that can reach from just an average

whistle to being as loud as Blue Whales are.

Blue whales are the larg-est animals ever known to have lived on Earth. The adult blue whale can reach about 100 feet. Blue whales can travel up to 20mph in pairs, but when alone they go slower like 10mph. Blue whales don't really eat much they would eat mostly small little fish or krill.

The whistle can be very loud when in land and underwater. Scientist will study a whale to see how they call each other and will try to remake the same thing but added into a whistle. The whistle will reach 750 miles on land and in sea, 1000 miles. The way people will hear is by them getting a small dot on their radar and will tell that it's someone stranded. If you think you're lost it it will float up and glow really bright like a small flashlight.

The problem is people getting lost at sea, held hostage, someone lost, or just use it as an average whistle. The percentage of surviving a plane crash is surprisingly really good.

You have a better chance of surviving if you sit near the rear part of the plane.when stranded the best thing to do is to stay calm at all times. Look out for boats or planes and find a way to write something on the floor so you can catch the attention of someone looking for you.

Remember now if you get lost somewhere you can have this whistle and can call some-

If the flight 370 had these whistles we wouldve already found them

one. The whistle can be heard by someone about 750-1000 miles away. Now, if Flight 370, the Malaysian Airline had these whistles, then we could have already found them. That's why we need these whistles, to rescue people and bring them back to their loved ones. The reason I chose this animal was because a Blue Whale has a really loud call.

Laryngeal Sac inflater

U-Fold
vocal fold
Homolog
(Yellow)

Barrel Cactus Building

by Nate Richards

You are in a blazing desert and you spot a clump of short round cacti, a lizard runs up and flicks its tongue on the cactus pulp. Drops of water come out and you decide to investigate. What you see is the Ferrocactus Cylindraceus or more commonly known as the barrel cactus.

The barrel cactus stores and purifies

> *The yellow/red flowers and yellow fruit always grow at the top of the plant.*

water. Also, the larger plants lean to the southwest. Many barrel cactus lean to the south so that a minimum of body surface is exposed to the drying effect of the midday sun. The barrel cactus is one of the largest cactus species in North America. The barrel cactus' spines can vary from straight to hooked. Some of the barrel cacti identifying features such as the fishhook barrel cactus

(Ferocactus wislizenii) can be identified by its thick (2 foot diameter), barrel shaped body and long hooked spines.

The yellow/red flowers and yellow fruit always grow at the top of the plant. Fishhook barrel cacti grow along desert washes and gravelly bajadas. It is less likely to occur on valley floors or rocky slopes. This species of barrel cactus is found in south-central Arizona and northern Sonora, Mexico. There are scattered populations in southern New Mexico and western Texas. Its lifespan is 50 to 100 years.

My inspiration came from my love of the state California to make a alternate way to use water to save us from a drought. I had tried desalination, but then we would have to use large amounts of seawater. This way we use the water that the rain gives us.

My building works by opening the funnel gate then the rain water pours in.

My building works by opening the funnel gate then the rain water pours in. A separate tank holds 1 gallon of iodine measuring the water as it fills in, every 8 ounces a drop of iodine goes in. A testing facility nearby test the amounts of water coming in and adds drop of iodine when needed.

In conclusion, by using adaptations of the barrel cactus, my design can help stop droughts and save lives. A device like this can also help third world countries by collecting and purifying their water.

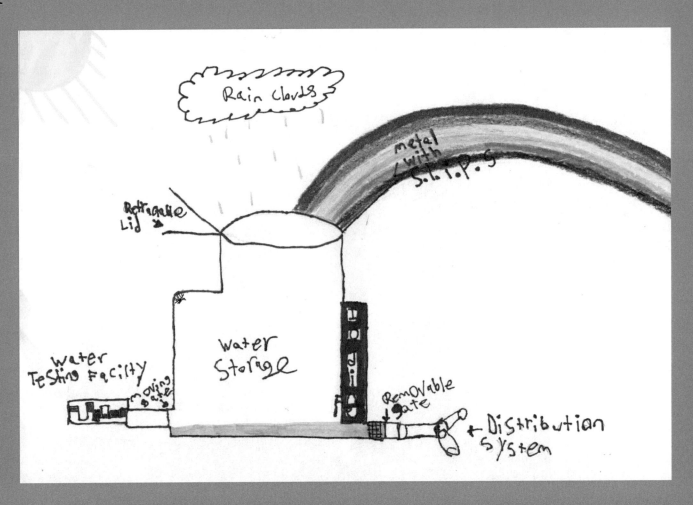

The Turtle Dome

by Alani Bayha

San Diego is still struggling with homelessness. Several people are living on the streets. Some of them have mental illnesses and others are veterans slowly failing to find hope for a better future. The solution is one easy thing, find houses, then we can give support, health care, and job openings (uwsd.org). But before that, protection against the elements is needed. If you're outside and it's raining, you probably don't think about it as a big issue. But if you're homeless and in the rain your only refuge is under a bridge or in a cardboard box. We need to change that. My solution is, a lightweight portable shelter.

I decided to do some research on tortoise shells to find a convenient shape for my design. I found that the top piece of a tortoise shell is called the Carapace. The bottom part of the shell is called the Plastron (kids.nationalgeographic.com). Also, the hard shell shields them from harm. Unfortunately, the shell does not cover their legs, belly or head. But the tortoise can pull back the legs as well as its head into the shell for safety (a-z-animals.com). My design will be a little different, because humans have a different body shape than tortoises, the dome will be able to slide in and out so it's easier to carry.

The dome will be made out of an aluminum alloy. Aluminum is a delicate, featherweight, fire-resistant, moldable metal that lets electricity flow through it. It bounces back light and heat very well, and it also won't rust. You can put aluminum together with other materials to create an aluminum alloy (alloys are a mixture of metals or other materials to create a new substance with enhanced properties). I plan to use an alloy of copper and aluminum to create my metal dome. Why copper? Because aluminum alone is delicate, and adding copper will make it more durable, tough, useful in high or low temperatures, and keep it lightweight (explainthatstuff.com). As you can see in the picture, the dome is made up of multiple curved parts bolted together so it can slide open and closed. To keep the heat in, I based my design off of a thermos. A thermos is a container you use to keep warm foods warm and chilled foods cool. A thermos does this by having two walls with no air between them, this creates a vacuum. It works by not letting heat get away (wonderopolis.org). This technique will be applied inside of each of the sliding metal plates of the dome.

Avoiding the elements is easy for us because we can take refuge in our homes, work buildings, restaurants etc.. What if you didn't have those options? Well that's what homeless people have to go through every day.

It is amazing how many people are homeless. One night, on January 2015, 564,708 people were homeless in the United States! Out of that, 206,286 were people in families! And 358,422 people were alone (endhomelessness.org)! Avoiding the elements is easy for us because we can take refuge in our homes, work buildings, restaurants etc.. What if you didn't have those options? Well that's what homeless people have to go through every day. It's tough to be homeless and some of the ways it's caused are divorce and mental illness. In the situation of a divorce, one of you gets the home while the other is left houseless. It gets worse if you (or your spouse) is jobless, than you (or they) would have a rough time getting a job or paying child support. Having a mental illness is no walk in the park either. You can become homeless by getting a costly remedy and aren't able to afford a home, or you don't take medication and lose your job because of that (povertyliving.com).

In conclusion, I copied the shape of a tortoise shell to create a portable dome made out of an aluminum and copper alloy for the homeless. They can use this to sleep under and keep warm or as an umbrella. It can slide together to make it easier to travel with too.

29938304R00062

Made in the USA
San Bernardino, CA
02 February 2016